THE
SHAPE
OF THE
WORLD

THE SHAPE OF THE WORLD

RAND McNALLY

CHICAGO/NEW YORK/SAN FRANCISCO

The endpapers *show details from a Dutch world map of the 17th century – the golden age of cartography, when maps became true works of art.* Front *Neptune with his trident bestrides the oceans. The legend on the left records the daring feats of Ferdinand Magellan, the renegade Portuguese sea captain who first sailed west around the southern tip of America and across the Pacific to Asia.* Back *The King of Spain surges across the Atlantic Ocean to visit his empire in central and south America. At the top American Indians spear whales off the coast of Nova Scotia.* **The frontispiece** *is a satellite photograph of North America, revealing the dramatic curvature of the continental surface.*

Published by Rand McNally in 1991 in the USA

© Granada Television Ltd
Simon Berthon
Andrew Robinson
All rights reserved

Library of Congress Catalog Card Number: 90–053291
ISBN: 0-528-83419-3

"The Shape of the World" is presented on the PBS network by Thirteen/WNET New York.
PBS television series funded by IBM.

Printed in Hong Kong

CONTENTS

INTRODUCTION

by Simon Berthon

In 1985, General Pete Thuillier, a retired officer of the British Army, told me the wonderful story of a small band of richly contrasting characters who had been responsible for an epic if apparently bizarre enterprise in the first half of the nineteenth century. They had set out to map the entire subcontinent of India with the utmost precision from the bottom to the top. Their amazing ambition was rewarded 50 years later with the discovery and measurement of the world's highest mountain, Mount Everest. They had worked in atrocious and often hostile conditions and many had died on the way; but their feat became one of the greatest if least appreciated achievements of nineteenth century science.

Anyone who has seen Pete Thuillier's appearance in the fourth program of our television series, THE SHAPE OF THE WORLD, will understand why I was instantly captivated. His tale had all the ingredients for a splendid film: an original subject, colourful characters, spectacular scenery, and a tremendous ambition realized against the odds. Six years later his inspiring idea had grown into six television programmes, covering not just India but the whole world, and this beautifully illustrated book.

On one level the story of the Great Trigonometrical Survey of India was about obsessive individuals who, in a genuinely heroic and selfless way, drove themselves towards a purely scientific goal. But it was much more. Today most of us would no doubt view maps simply as things that enable us to move around, and a modern atlas as an objective and accurate view of the world. But during the thousands of years in which people have sought to make sense of the world around them, there has been no such thing as a neutral map. In India, for example, the map was a vital tool of empire. Imperialists required an accurate picture of their subject land in order to rule it effectively. As our project expanded from India to the rest of the world it became clear that maps were a mirror of history, their makers driven by fundamental forces ranging from science and religion to power and greed.

In the beginning was a simple and devastating question. How could anyone rooted on earth, however ingenious, begin to work out what the world actually looked like? Today we are so spoiled by those fantastic luminous pictures from space of our globe floating unsupported in the universe that it is hard to imagine what it must have been like a few thousand years ago, when the first mystics and scientists set out to understand the world. They had none of the technology which we take for granted: no satellites or airplanes from which to look down on the world, and no fast and safe transportation with which to discover what lay beyond the horizon.

This tantalizing mystery was actually enhanced at the outset by the absence of technology. For early societies, unencumbered by any awkward facts which might stand in the way of their illusions, had the licence to invent all sorts of intriguing ideas. For some reason the Indians seem to have had the most attractive imaginations. A vision of the world supported by a tortoise which lived on the backs of four elephants was one charming result.

The understandable urge to find something on which the world could rest cut across many cultures and it required the rejection of religious imagination and the introduction of science by the Greeks to get rid of it. Despite the weird constructions of the early Christian, Cosmas Indicopleustes, and his fervent attempts to ridicule the pagan science of those like Claudius Ptolemy, Greek science was ultimately irreversible. Ptolemy lasted rather longer than Cosmas.

INTRODUCTION

If science and religion were the competing forces which first grappled with the shape of the world as a whole, baser urges propelled the quest to map what lay on the Earth. The Egyptian Pharaohs surveyed the fields of the Nile basin in order to charge the correct taxes; the warring states of ancient China used maps when fighting each other; Alexander the Great mapped his expanding empire and in the process hugely increased Greek knowledge of the world; the Romans wanted maps above all for the administration of their conquests.

As Europeans hungered for new lands in the Age of Discovery, sea charts became vital documents of state. Knowledge of the oceans and coasts was highly prized, the key to wealth and empire. Indeed you can chart the rise and fall of Portugal between 1400 and 1600 by the activities of its navigators and mapmakers. What began with the achievements of Henry the Navigator's sailors degenerated into the sale of Portuguese secrets by the chartmakers, Diogo and Andre Homem, and the tawdry exploits of the pilot, Simon Fernandes, who traded his skills to Elizabethan buccaneers like Sir Walter Raleigh.

The settlement of the New World gave maps and surveying a vital new relevance. While Indians lived communally, Europeans had an individual and possessive view of the land which was formalized in maps. They became the seals which conferred a spurious legitimacy on the European takeover of America. Some fascinating recent research demonstrates how the earliest European maps acknowledged Indian names and territory, and how these were then quite literally wiped off the face of the map.

Despite the growing need for maps, the limits of technology meant that less than one eighth of the world had been accurately surveyed by the end of the nineteenth century. Then, with the help first of gas balloons and culminating with satellites and computers, the final mysteries of the Earth's shape, both visible and invisible, were revealed. With these breakthroughs, mapping seemed to be driven once again by science rather than power. The results initially appeared beneficial: for example the opportunity to resettle landless peasants in the Amazon rainforest. But today the new high-tech maps are double-edged weapons. What enables us to exploit the Earth's resources also empowers us to destroy it. Will the continent which lies beneath the ice cap of Antarctica be allowed to remain an untouched wilderness, or must the mapmakers' skills lead to the disgorging of its oil and minerals?

The television series and this book are indebted to many people throughout the world. I am deeply grateful to all of them, and hope that they will understand if, in this brief foreword, I restrict myself to mentioning those colleagues who have helped to realize the project. Mike Scott, then program controller at Granada Television, was hugely encouraging and influential in the early stages, and Rod Caird, head of features, was instrumental in making it all happen. The backing of IBM was the *sine qua non*, and I would like to thank all at IBM and their consultants not just for their sponsorship but also for their part in forging a happy relationship.

It is, of course, one's colleagues on the production team who make or break a project like this and here I have been very lucky. They are Eleanor Burnikell, Allison Denyer, Janice Finch, Claudia Josephs, Emma Laybourne, Lynne Marriott, Mark Anderson, Alex Connock, Patrick Lau and Peter Swain. There are no broken bones nor even smashed coffee cups to be seen in the office and I hope that, despite arduous times, they too found it fun. To those of them with whom I particularly discussed the book and who were so generous with their thoughts I am most grateful. Finally many thanks to my editor, Richard Dawes.

1

HEAVEN AND EARTH

In December 1968 the three American astronauts of Apollo 8 achieved one of humankind's most improbable ambitions. As their spaceship left the cocoon of Earth's gravity, hurtling them into outer space on their epic journey to circumnavigate the moon, they became the first human beings to see the shape of the world with their own eyes. The Christmas present they sent home was a series of photographs and television pictures. And that vision of Earth, a ghostly sphere with luminous green lands and indigo seas veiled by swirling bands of cloud, became one of the most compelling images of our age.

William Anders was on board Apollo 8. All of his technical and geological training, years of dedication and commitment, had been focused on the moon; this was a lunar program. The closer he drew to his target, the more barren and lifeless it seemed. He gazed back at the Earth receding into the distance, shrinking down into its real solar and universal perspective, suspended against the great black backdrop of space. On the fourth orbit of the moon the crew emerged from the dark side and rotated their spacecraft. Suddenly before them lay an unimaginably spectacular sight, an Earth-rise. The image imprinted itself indelibly on Anders. "For the first time," he recalls, "we saw the beautiful orb of our planet coming up over this relatively stark, inhospitable lunar horizon and it brought back to me that indeed even though our flight was focused on the moon, it was really the Earth that was the most important to us." It was Christmastime. The friendly object a quarter of a million miles (400,000 km) away reminded Anders of a very beautiful and fragile Christmas-tree ornament, an object insignificant in the enormous cosmos and needing to be handled with care.

This awesome vision, the idea of the Earth as a living organism, became the most potent symbol of a dawning ecological awareness. But Anders and his fellow astronauts, Frank Borman and James Lovell, were also the first witnesses of a more basic truth. Until Apollo 8 our physical idea of the world had been supposition, the product of complicated mathematical equations and the piecing together of photographs from satellites and spacecraft still closely hugged by terrestrial gravity. For many, perhaps, a simple act of faith was involved. Now the human eye could see that the Earth really was round, a multicolored sphere spinning in the cosmos, the final answer to the puzzle which had gripped the human mind for thousands of years. Today the image of the Earth softly glowing in space is so vivid and familiar that it is hardly possible to imagine that things were ever different. Television news bulletins start with an image of the globe, World Cup soccer logos are based on it, advertisements exploit it, airline symbols would be lost without it. We talk of global warming, the global village, the global arms race, the global network and global communications.

It was only two and a half thousand years ago, a mere instant in the life of the Earth, that a few scientists and philosophers in Greece conceived the idea that the world was a sphere. The understanding of that sphere and what lay on it followed slowly and painfully, with many interruptions and reversals. From the very beginning the road to enlightenment had been marked with deep craters of superstition, myth and religion. Ancient civilizations existed in relative isolation, without the sophisticated communications that we take for granted today. So how could people whose perception was bounded on all sides by the horizon and by the skies above begin to create their own picture of the world? The first answers lay in their imagination. Above all they were responses to two basic, primeval questions: Where am I? and What happens to me when I die? The mental constructions of the universe devised to deal with these mysteries stemmed from two sources. On the one hand there was the empirical knowledge of what could be seen; on the other hand was the need to secure salvation by having recourse to the supernatural.

Early ideas were colored by a number of prejudices, all of them understandable and comforting. For a start there was the reassuring fact that the Earth stood still. If it was moving, we would feel ourselves moving. It must be flat, otherwise we would fall off. It was also immense, stretching without limit into unknown oceans and lands. The world was clearly at the center of the universe. You had only to look up into the skies to see the sun, the moon and the stars revolving around it. Its centrality stemmed from a basic human response: here I am, and I stand at the heart of all that is around me.

Almost every society, whether Chinese, Aztec or Iroquois Indian, viewed itself as at the center of the world. When Europeans had discovered the full extent of North and South America, Australia and Antarctica, any schoolchild brought up on Mercator's map would have noticed that Europe was placed firmly in the middle. Long after the basic order of the universe had been revealed, this attraction of the center, known as the *omphalos syndrome* from the Greek word for navel, persisted in people's vision of their place on Earth. Even today *New Yorker* magazine's map of the world is only half joking when it shows that city occupying much of the planet's surface and everywhere else receding into the distance.

In ancient Egypt the center of life was the Nile. The Egyptians' geography was shaped by the striking contrast between the rich soil of the fields bordering the river, known as the Black Land, which was Egypt, and the Red Land, the unrelieved mountainous landscape that lay beyond. The Nile divided this landmass right down the middle and then swept around it in the Great Circular Ocean. Egyptian Creation myths depended on this great profusion of primeval water, represented by the god Nun, in which the world floated, divided by a firmament from the waters above, the source of floods and rain. Above this bowl of earth with its ring of mountains, the sky was held up by four supports, one at each corner. In some representations these supports were poles or forked branches, in others four great mountains towering above all else. Where did the sun, the second dominant force of life, fit in? Here the imagination had free rein. The sun was a hawk which rose serenely from the eastern

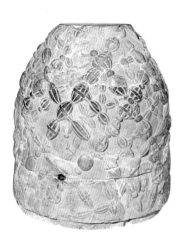

The omphalos of Delphi, the sacred stone which marked the center or navel of the ancient Greek universe. Early civilizations throughout the world were bound by the omphalos syndrome, each firmly believing that it lay at the center of everything.

hills to climb magnificently into the heavens; it was a great celestial cow; at dusk it was swallowed by the goddess Nut through whose body it passed, allowing her to give birth to it again each morning; it was pushed across the sky by an enormous beetle. Most such images derived from an attempt to relate visible experiences to the fantastic event which daily took place above. The notion of the beetle, for example, was inspired by the common scarab beetle's habit of rolling its ball of dung before consuming it in the darkness of its lair.

The ancient Egyptians also exhibited an impulse which has always dominated visions of the world. People illustrated or made maps for their own innermost reasons. In Egypt the Pyramids are the massive evidence of that civilization's preoccupation with the afterlife. The location of heaven and what lay there were matters of supreme importance. Again, the Egyptians' image of heaven was drawn from everyday earthly experiences. The life ahead, so beautifully illustrated in tomb paintings, became an idealization of the life made possible by the Nile. The Egyptian heaven was depicted as the Field of Reeds where men and women could plough, pull flax, sow their seeds, and reap the grain. In the bottom of coffins, paintings showed the two ways to the afterlife, by water and by land—possibly the world's first route maps.

Unlike Egypt, Mesopotamia, the second center of early Near Eastern civilization, left us a world map. The impression on a clay tablet is tiny—5 × 3 in

Ruined temples at Luxor as seen from the Nile in a nineteenth-century painting. The Egyptians' monuments embodied their religious conception of the world, with the cities of the living facing the cities of the dead across the sacred river. On a more mundane level, the fields along the Nile were precisely mapped after floods each year to help the Pharaohs collect their taxes efficiently.

(12.5 × 8 cm)—the size of a hand. There are two lines running down the center of the flat, round Earth and these probably represent the Tigris and Euphrates rivers. In a central circle are Babylon and other important cities. Surrounding these is the circular Bitter River beyond which lived all sorts of legendary beasts. This crude picture of the world seems a disappointing memento of a civilization which was in many other respects so brilliant. The Babylonians were, for example, superb astronomers. Their figures on the motions of the sun and moon reveal only three times the margin of error of nineteenth-century astronomers armed with powerful telescopes. Yet when it came to picturing the world and the structure of the universe, science was thrown to the winds, and experience surrendered to myth. For example, the flourishing plants of the river basin, watered from below, led the Babylonians to conclude that there must be an immense body of subterranean water, called the Apsu, provided by its patron god, Enki or Ea, to whom great monuments were constructed. Above this was the flat disc of earth, the domain of another god, Enlil. Higher still in heaven, despite the Babylonians' sophisticated understanding of astronomy, room was found for a succession of palaces in which the gods held court just like earthly kings and queens. However, the underworld, where the dead went, was taken to be a dry and dusty place, contradicting the notion of a great water below. The solution was that the dead should have to cross the water in order to reach their final resting place.

The Babylonians devised their complicated layered universe to accommodate their myths and gods rather than by their astronomical observations. It provided reassurance and comfort, whereas the uncertainties of science could have led to anxiety and disorientation. Like the Egyptians, they preferred to stay in the great oyster of the world, surrounded by life-giving water and the domains of the gods. The impasse could be resolved only by a new attitude towards the world, by asking a different set of questions. It was to be the Greeks who first grappled with the reality of the cosmos. It was a slow process, for the Greeks too had to leave behind their gods. The first glimpse we have of their world view is an entrancing description by Homer in the *Iliad*. Patroclus, the bosom friend of the Greek prince Achilles, has been killed in battle by the Trojans. He was wearing Achilles' armor, which has been taken from his dead body and is now being worn by his Trojan vanquisher, Hector. Achilles, who has refused to fight after quarrelling with King Agamemnon, the imperial overlord of Greece, is torn by remorse. However, spurred by the desire for revenge, he decides to rejoin the battle.

Achilles' mother, the goddess Thetis, hears her son's distant cries of anguish and soon joins him at the Greek fleet encamped on the shores of Troy. She warns him that if he slays Hector in battle, he himself will die. But Achilles will not be deterred, so his mother decides that at the very least he must have a new set of armor. Thetis visits her fellow god, the Master Smith Hephaestus, from whom she commissions a helmet, a pair of fine greaves fitted with ankle-clasps, a cuirass and a shield. On the latter Hephaestus created a map of the world. Setting to work with his twenty bellows on imperishable bronze, the divine smith decorated the shield with designs of the Earth, Sky and Sea, the Sun, the full Moon and, as Homer describes it, the constellations with which the heavens are crowned, the Pleiades, the Hyades, great Orion and the Bear. Hephaestus then depicted two cities, contrasting scenes of weddings and banquets with those of trials and battles. He added vignettes of ploughing and reaping, cattle and vineyards, and finally a dancing floor crowded with youths wearing gold daggers on their silver belts and maidens with lovely garlands on their heads. "Finally," Homer tells us, "round the very rim of the wonderful shield, he put

The Pyramids at dawn. The preoccupation of the Egyptians with death dominates the landscape, for the pyramids were monumental burial chambers. They also reflect an Egyptian idea of the world in which the sky was held up by four enormous supports, planted at each corner of a flat, rectangular Earth. The sun was portrayed in many delightful forms, from a celestial cow to a colourful scarab beetle.

This tiny clay tablet, the size of a hand, shows the oldest surviving world map. It was made two and a half thousand years ago by the Babylonians. Earth is a flat disc with Babylon at the center and the Tigris and Euphrates rivers prominent. Around the edge was the ocean surrounding the world.

The Field of Reeds: the Egyptian heaven. The location of the afterlife and the routes to it were matters of enormous importance to the Egyptians, so the life ahead was mapped in tomb paintings like these. In the Field of Reeds, men and women could plough, pull flax, sow their seeds and reap the grain, all the activities which made life on Earth comfortable.

the mighty stream of Ocean'' – a vast river encircling the world.

Hephaestus' decorated shield was hardly a scientific map but it anticipated many of the concerns which were to fascinate philosophers, scientists and geographers. Which parts of the world were inhabited and who lived there? What lay at the edges of the world? It foreshadowed maps as keys to understanding human distribution. Yet it failed to answer the basic questions about the world. In particular, his depiction of the Earth as a flat disc resting on water provided no

The world as reconstructed from clues contained in Homer's Iliad and Odyssey. The known world is very small, and centered on the eastern Mediterranean. Greece, from where Agamemnon and his army set out, is shown in detail, as is the western coast of Asia Minor, where lay their target, Troy. The waters of what Homer called "the mighty stream of Ocean" surround everything.

solution to the ancient search for the layers of the universe which, it was thought, must exist in order to hold each other up. This search explains why the Egyptians postulated pillars and the Babylonians a great sea below. If such devices were not present, would not the sky and the Earth itself simply fall away and disappear?

The question will always trouble some. One of the greatest physicists of the late twentieth century, Professor Stephen Hawking, the master of space-time curves, quasars and black holes, tells a story in his best-seller *A Brief History of Time* about a well-known scientist, allegedly Bertrand Russell, who gave a public lecture on astronomy. He described the basics of the universe; how the Earth orbits around the sun and the sun then orbits around the center of the galaxy. "At the end of the lecture," writes Hawking, "a little old lady at the back of the room got up and said: 'What you have told us is rubbish. The world is really a flat plate supported on the back of a giant tortoise.' The scientist gave a superior smile before replying, 'What is the tortoise standing on?' 'You're very clever, young man, very clever,' said the old lady. 'But it's turtles all the way down!'" In fact giant tortoises or turtles have often occurred in representations of the world. But for most of us they never quite provide an adequate explanation.

The riddle of the world's physical structure prompted a search for its true shape and location in the universe. The breakthrough came in the town of Miletus in the sixth century BC. Miletus, now a desolate spot on the west coast of Turkey, established nearly one hundred colonies stretching from the shores of the Mediterranean through the Hellespont to the Black Sea. Traders and seafarers from all over the inhabited world then known to the Greeks sailed in and out of its harbor dropping and taking on cargoes of olive oil, grain, perfume and wine. This pleasure in life, this prosperity which has so often deflected men from finding consolation in the gods to pursue material comforts instead, may have been the spur to the scientific revolution which took place in Miletus. Here, for the first time, men tried to explain the universe without resorting to the facile solution that everything must be controlled by the gods. The Italian philosopher, Luciano de Crescenzo, sums up the contribution of the Milesian philosophers of over two and a half thousand years ago: "Previously, if it rained, it was Zeus' fault. If there was an earthquake it was Poseidon's fault. With the Milesians, people start to think that all events have a physical cause. They were the first scientists."

Asking the right questions was of course no guarantee of getting the right answers and some of the early theories seemed to make little advance on Egyptian and Babylonian fantasy. But they were arrived at in a totally novel way. The first beacon of enlightened enquiry was the astronomer Thales. He made a name for himself by correctly predicting an eclipse of the sun, an enviable gift in any ancient society. Like Homer, he believed that the Earth was a flat disc resting on water, but he came to this conclusion only after a diligent search for the physical causes of the universe. He decided at length that the one basic element must be water, because everything was born from it, even the air, which is evaporated water. Thales and his fellow thinkers such as Anaximander and Anaximenes set off an intellectual explosion. The Milesians were edging towards new explanations. Anaximander concluded that the universe was infinite in time and space. More significantly he freed the world from the traditional requirement that it must have something to rest on—an enormous advance on Thales' foundation of water. Instead he allowed the world, which he probably believed to be a cylindrical column, to hang freely in space, floating without support. According to Anaximander, it did not fall because, being at the center, there was no particular direction in which the column should lean.

Those who, like Anaximander, speculated on these mysterious questions were philosophers, not experimental physicists. Their ultimate desire was to construct perfect working models of the universe which would accurately reflect the undeniable movements in the heavens. So Anaximander's ultimate creation was a model of a great series of fire wheels circling the Earth in neat, mechanical order. The idea of a circle was a powerful one and it would shortly be incorporated into the conception of the Earth itself. Unfortunately, no one will ever know exactly who first came up with the idea of the world as a sphere rather than a flat disc. The advent of the most likely candidate for that honor is exquisitely described by Arthur Koestler in *The Sleepwalkers*: "The sixth century scene evokes the image of an orchestra expectantly tuning up, each player absorbed in his own instrument only, deaf to the caterwauling of the others. Then there is a dramatic silence, the conductor enters the stage, raps three times with his baton, and harmony emerges from the chaos. The maestro is Pythagoras of Samos."

Pythagoras was a pupil of Anaximander, the master of physical matter and, by contrast, of Pherecydes, a mystic who taught the transmigration of souls. Combining these two influences he developed a range of ideas which were awe-inspiring in their breadth, for they comprised mathematics and medicine, geometry and music, mysticism and science. Pythagoras was born in Samos for whose enlightened tyrant Polycrates he carried out diplomatic missions during his extensive travels in Egypt

The theater at Miletus in Turkey is the only significant remnant of a city which in the sixth century BC was a great metropolis of the civilized world. But even this theater is Roman, built over the Greek original. It was in Miletus that for the first time philosophers like Thales and Anaximander began to think in a scientific way about the world and break the grip of mythology.

and Asia Minor. Around 530 BC he moved to Kroton (now Crotone), the second largest Greek town in southern Italy. It was in this town that he founded the Pythagorean Brotherhood, which for nearly thirty years dominated the thinking of much of Greater Greece. Modern Crotone's only relics of that extraordinary period are a few terra-cotta statues in the local museum. Even more disappointingly, none of Pythagoras' writings survive. Yet his influence is timeless. He lived to perhaps over ninety and into this long life he packed, according to a later disciple, ''all things that are contained in ten, even in twenty, generations of men.'' Among the discoveries made by Pythagoras, his Theory is probably the best known. This states that the square of the hypotenuse of a right-angled triangle equals the sum of the squares of the other two sides. Another of Pythagoras' discoveries was the relationship of musical notes, the revelation that a note's pitch depends on the length of string which produces it, progressing in intervals of numerically simple ratios, octaves, fifths, fourths and so on.

But the most exciting application of Pythagoras' notions of harmony was to the universe itself. The sphere became the perfect shape of the cosmos with the spherical Earth itself at the center, supported by nothing and simply surrounded by air. The sun, moon and planets revolved in concentric circles, each locked into a giant sphere. Their movement caused them to hum in the air, and the pitch of their note depended on the length of their orbit and could be related to musical intervals. In this cosmic musical scale the interval between the Earth and the moon was a tone; between the moon and Mercury, a semitone, with further intervals all the way out to the sphere of the fixed stars. This mixture of Pythagorean mathematics and mysticism, the Harmony of the Spheres, never lost its fascination, inspiring many lines of verse, like these, for example, of John Dryden:

> *From harmony, from heavenly harmony,*
> *This universal frame began:*
> *Where nature underneath a heap*
> *Of jarring atoms lay*
> *And could not heave her dead.*
> *The tuneful voice we heard from high:*
> *Arise, ye more than dead.*

Ordinary mortals were unable to hear the celestial harmony because they had been bathed in it since birth or because they were too imperfect to tune in to it.

Pythagoras' follower Philolaus liberated the spherical Earth from its awkward immobility. If it did not move, it meant that all the heavenly bodies from the nearest to the furthest must orbit the Earth every day at enormous speed. But if, as the Pythagoreans came to believe, the Earth revolved on its axis once every 24 hours, enabling us to take a sweeping look across the day and night skies as we spun in our slow pirouette, astronomical observations would become more coherent. There was an even more exciting idea. Perhaps the Earth, as well as spinning, had its own orbit. To account for this a central fire, quite different from the sun, was conceived as the new focal point for the universe. We could not see it because it was always on the opposite side of the world. Nor could we see the Antichthon, or counter-Earth, which revolved between the Earth and the central fire and served the important function of protecting the other side of the world from the fire's heat. The Antichthon brought to ten the number of bodies—Earth, sun, moon, planets and fixed stars—thought to revolve around the fire. Ten was a magical number for the Pythagoreans.

These ideas nearly led to the unravelling of the puzzles of the universe. The Pythagoreans were seeking a solution to the problem that the observed motions of the planets simply did not suggest a pleasing series of circles moving serenely around a stationary Earth at the center. Instead the planets wandered along, sometimes speeding up, sometimes slowing down, and in the case of Venus alternately approaching and receding from us. Later Pythagoreans broadened their outlook further to take account of these inconsistencies. Heracleides held that Venus and Mercury orbited the sun, which along with everything else, then circled the Earth. Most innovative of all was Aristarchus of Samos. Towards the end of the fourth century BC, nearly two thousand years before Copernicus, he produced a solution that accommodated all the vagaries of the heavens. His own writings are lost but his younger contemporary, the more famous Archimedes, confirmed that Aristarchus "supposed that the fixed stars and the sun are immovable, but that the Earth is carried round the sun in a circle." Aristarchus was right, yet his idea died with him.

The mainstream Greek tradition could not cope with the demeaning possibility that the Earth was less than pivotal to the whole scheme of things. Observations had to fit in with the notion of the Earth-centered wheel even if planetary circles were imperfect. To illustrate their complicated cosmology the Greeks invented celestial globes. One, albeit a Roman copy of a Greek original, survives in Naples. Known as the Farnese globe, after the family to whose collection it belonged, this magnificent specimen rests on the muscular marble shoulders of Atlas, who understandably stoops under the crushing burden of the universe. Although the Greeks had left behind this world view hundreds of years before, their statues continued to evoke the Homeric idea of the mythical giant holding up the heavens.

The Farnese Globe, now in the Archaeological Museum in Naples: a second-century AD Roman development of a fourth-century BC Greek original. This is a celestial globe, showing the universe as if we were looking at it from the outside, with the Earth at the center of the spherical heavens.

A six-day-old moon. The apparently circular movements of the moon, sun and stars led the Greeks to believe that the Earth was a perfect sphere. One piece of evidence cited by the Greeks was the curved shadow cast during lunar or solar eclipses.

The Farnese globe shows the universe as if we were looking at it from the outside. Forty-three constellations are carved on its surface, including Cancer the crab, Taurus the bull and the other astrological signs. In its detail it is a beautiful and scientific picture of the night sky, yet there is the impenetrable anomaly that somehow everything must wheel around this one small planet stuck in the middle. It was an astronomical cul de sac, a complex machine of wheels within wheels in which planets had to be given their own mini orbits or epicycles to take account of their inconsistencies. In short, classical ingenuity was increasingly applied to oiling a creaking mechanism rather than searching for a clean new engine.

Understanding of the Earth itself would also be caught in the crossfire between, on the one hand, the desire for symmetry and the ideal, and, on the other hand, the evidence of everyday experience. Whether or not Pythagoras himself first suggested that the Earth was a sphere (the other main candidate for the honor is Parmenides, another Greek expatriate in southern Italy; some even believe that Anaximander's Earth was a sphere rather than a cylinder), the idea began to be firmly established in the Greek mind. Confirmation appears in Plato's *Phaedo*, written in about 380 BC. There Socrates describes the Earth: "I've been convinced that if it is round and in the center of the heaven, it needs neither air nor any other force to prevent its falling, but the uniformity of the heaven in every direction with itself is enough to support it, together with the equilibrium of the Earth itself."

Shortly afterwards the seal of respectability was given to the theory of the Earth's sphericity by Aristotle, the teacher of Alexander the Great and the dominating intellectual force of the fourth century BC. Aristotle relied on observation. For example, the shadow of the Earth in a lunar eclipse is incontrovertibly round. This effect could also have been caused by a cylinder, but Aristotle had supporting evidence. By this time, travellers' reports recorded that the further north or south you went, the lower in the night sky some stars appeared to be. This could only happen if constant progress round a curve changed the angle of sight. Finally, Aristotle noticed that ships disappeared bow-first over the horizon.

It was one thing to work out that the world was a sphere, quite another to determine what was on it. This pursuit turned into one of the great games of the ancient world, full of delightful disputes and nitpicking, claims and counter-claims. At the root of it all was once again Homer. Later Greek writers indulged in drawn-out and acerbic arguments about whether Homer, a poet after all and not a philosopher or scientist (if indeed a single person at all), could be expected to have paid any attention to fact as he unfurled his story of the siege of Troy and the wanderings of Odysseus on his arduous journey home to Ithaca. However, there is much evidence to suggest that Odysseus' encounters with bizarre races such as the Cyclopes and with the perils of Scylla and Charybdis, his visit to the land of the lotus-eaters, and the geographical descriptions which accompanied these episodes, did indeed relate to identifiable places on the coasts of Asia Minor, North Africa and Greece. But while the distances may have been vast to Odysseus as his single-sailed boat was buffeted by gales and driven onto treacherous rocks, the scope of his voyage, as we can complacently observe today, was small compared with the true extent of the world.

It is the very confines of this restricted world, hemmed in by the Mediterranean, which give the story its excitement. In western Turkey, Greece, or southern Italy nearly three thousand years ago, travel was slow, difficult, and dangerous, and information scarce. To the north and east the land seemed to stretch forever. To the south, across the Mediterranean, there was another landmass, usually called Libya, in which people known as Ethiopians were thought to live. (At the very beginning of the *Odyssey*, Homer made a tantalizing reference to "the distant Ethiopians, the farthest outposts of mankind, half of whom live where the sun goes down, and half where he rises.") At the western tip of this Mediterranean world huge rocks called the Pillars of Hercules towered over a strait (today called the Strait of Gibraltar) which led into a seemingly endless sea called variously the Ocean Sea or the Atlantic. This was the world picture which confronted the Greeks. How they filled in the pieces is one of history's great detective stories.

Once again the process started with the Milesians. Later writers stated that Anaximander of Miletus was the first person to draw a map of the world. Apparently he also made a globe. He is thought to have invented the gnomon, a straight, vertically held rod in some cases as long as 35 ft (11 m), on which the Greeks came to rely for measurements of latitude. Its operation was simple: experience gave a knowledge of the precise overhead location of the sun on particular days of the year, and on these days at midday the shadow cast by a gnomon would give an angle from which the latitudinal distance of any place could be calculated. It was a rough-and-ready calculation, made tricky by the size of a gnomon, which was too long for travellers to take everywhere they went.

Knowledge of the world therefore had to be built up from explorations and voyages, the tales of travellers and traders, and military conquests. The finest early description was by Herodotus. Born at Halicarnassus on the southwest coast of Asia Minor towards the beginning of the fifth century BC, Herodotus himself visited Egypt as far south as Aswan, Mesopotamia, Palestine, southern Russia and the northern fringe of Africa—in fact, most of the world known to the Greeks. He then retired to Italy to write his history. That is all we know of him; the rest must be imagined from the wit, acerbity and grandeur of his great book. "I cannot help laughing," wrote Herodotus, "at the absurdity of all the mapmakers—there are plenty of them—who show Ocean running like a river round a perfectly circular earth, with Asia and Europe of the same size." This was a frontal attack on the tradition that started with Achilles' shield and was taken up by Anaximander and successors such as Hecataeus, who is known to have drawn a map just like that described by Herodotus. In splendidly dogmatic language Herodotus set down his own description of the three known continents of Europe, Asia and Africa: "Europe is as long as the other two put together, and for breadth is not, in my opinion, even to be compared with them." But at least Herodotus was unwilling to be seduced by easy solutions such as the Ocean Sea when there was no evidence that there was anything but land to the north and east of Europe or beyond India.

One story of Herodotus seemed calculated to set the pulses of future generations racing. It concerned the Phoenicians, the remarkable seafarers who were an offshoot of the Semitic tribes of Canaan. They founded their nation in the west of today's Lebanon, around the towns of Tyre and Sidon. It is believed that from there spread the tentacles of a vast trading empire which straddled the Mediterranean long before Greeks, Egyptians or Persians dared to travel such distances by sea. Their sailors were called "red men," owing to their continuous exposure to the elements. They were military and mercantile mercenaries, fighting naval battles for Mediterranean nations. They shipped in a staggering variety of goods including molluscs, from which a purple dye for textiles was extracted, silver, iron, lead and tin. The latter probably came from Cornwall and the Scilly Isles and was collected by the Phoenicians on their northern trade voyages. They even traded amber, which was only found to the far north in the Baltic. No Phoenician accounts of this trade

A Phoenician ship of the seventh century BC, a bireme with two rows of oars. The Phoenicians were the great maritime traders of the ancient world, covering the Mediterranean and the coasts of Europe and northwest Africa. One tantalizing story related that the Phoenicians had circumnavigated Africa, an astonishing achievement in an oared vessel like this.

survive, so we do not know whether they went there to collect it or it was transported overland to trading stations at the head of the Adriatic.

Although the Phoenician records were lost, Herodotus' amazing story of their rounding of Africa remains. He writes that the Egyptian king Neco, who reigned from 616 to 600 BC, sent out a fleet of Phoenician ships to sail south from the Red Sea and then head west to round Africa and return to Egypt by way of the Western Ocean. Neco's motive is a grim reminder of those harsh times and of the vaulting ambition of monarchs. He had wanted a canal dug to connect the Red Sea and the Nile but many difficulties arose during the work and it was only after 120,000 laborers had died that he was finally persuaded by an oracle to abandon the project. The route round Africa was to be an alternative link from the eastern seas to the Mediterranean, predating the Portuguese quest for a sea passage to India and the Spice Islands of Indonesia by more than two thousand years.

The Phoenician fleet sailed into the Southern Ocean and, according to Herodotus, put in at some convenient spot on the East African coast, where they sowed a patch of ground and waited for the harvest. Having got in their grain they put to sea again and after two years returned through the Pillars of Hercules. In the third year they returned to Egypt. Herodotus was happy to believe in the possibility of rounding Africa, yet curiously he discounted the Phoenician account for the very reason which demonstrates that it may have been a historical fact. "These men made a statement which I do not believe myself," he wrote, "that as they sailed on a westerly course round the southern end of Libya, they had the sun on the right—to northward of them." This meant that the Phoenicians must have crossed into the southern hemisphere, which we now know would have been the case if they had indeed rounded the tip of Africa. To Herodotus this would have been an impossibility. Africa was a far smaller land, firmly contained within the northern hemisphere; the notion of it stretching thousands of miles to the south was inconceivable.

Herodotus nevertheless goes on to recount a gory story which supports the idea of an enormous continent. A certain Sataspes had been found guilty of rape and was on the point of being impaled by the Persian king Xerxes. Sataspes' mother intervened, suggesting that he be allotted the even more dreadful fate of making the long voyage round Africa, from west to east. Sataspes sailed south from the Pillars of Hercules. He continued for months and months but found that however far he sailed there was always farther to go and yet no sight of the tip of Africa. He returned home and tried to placate Xerxes with stories of pygmies who wore only palm leaves. The king was unimpressed and, as Sataspes had failed in his task, had him impaled after all.

From clues like these the Greeks might have reached the correct conclusion about Africa. But they were essentially philosophers rather than experimental scientists. The notion of a spherical world stirred in the Greek mind all sorts of enthusiasms for how that world could be most satisfactorily divided up. If the heavens could be divided into zones governed by musical intervals, could not the same symmetries be imposed on the world? Greek knowledge of the locations of the Tropics of Cancer and Capricorn made this enterprise easier. They divided the Earth into five rings. At the north and south were the Arctic and Antarctic rings, too cold to support human life. At the center was the equatorial or torrid zone where the sun scorched down at temperatures too high for man to survive. Sandwiched between these in the north and south were the temperate zones which alone were comfortable enough for human existence. These zones exerted an enormous power over both Greek and later

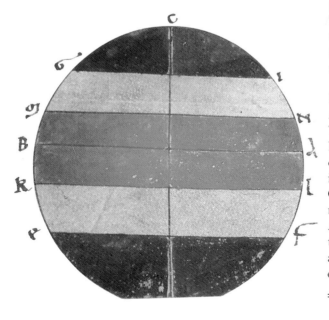

A passion for symmetry led the Greeks to divide the world into perfectly balanced climatic zones. In the middle around the equator was the torrid zone, too hot for human habitation. At the north and south were freezing zones which were too cold to support life. Between these were two habitable zones, but it was impossible to pass from one to the other because no one could survive the passage through the equatorial torrid zone.

thought. Their boundaries might be changed and a more flexible view might emerge of their nature, but their very existence debarred a belief in stories of humans sailing across the torrid zone into the southern half of the world.

At the beginning of the fourth century BC the Greeks' understanding of the world's composition was still restricted by a paucity of information. The resulting frustration was suggested by Socrates in Plato's *Phaedo*. The Earth, he said, "is of a vast size and we who dwell between the Phasis River and the Pillars of Hercules inhabit only a small part of it, living around the sea like ants or frogs around a marsh, and there are many others living elsewhere in many such places." All true, but where were the places and who were the others? Military conquest provided at least some of the answers. In the spring of 334 BC Alexander the Great crossed the Hellespont. Over the next decade, after a succession of victories over the Persian King Darius, he pursued his prodigious ambition of finding the ends of the inhabited world. His tutor and guide Aristotle had convinced him that such an aim was not unreasonable. The domain of man stretched from the Pillars of Hercules in the west to India in the east, and was completely surrounded by an ocean. Aristotle gave little credence to the idea of people living in landmasses on the far side of the world. To the contrary, the inhabited world was relatively small and could be encompassed by his royal master.

After his defeat of Darius at Gaugamela near the River Tigris in 331 BC, Alexander marched into Persia, destroyed Persepolis and took the Persian title of Great King. Over the next few years he penetrated into the Hindu Kush and the Punjab until finally his troops refused to venture beyond the Indus. Like poor Sataspes, who never managed to find the end of Africa, Alexander was thwarted in his desire to find the limits of Asia and perhaps declare himself the first universal monarch. In 323 BC he died of fever in Babylon at the age of thirty-two. If the brevity of his life was the retribution for such overweening ambition, Alexander's legacy was matchless. In the tradition of Greek monarchs and tyrants, the acquisition of knowledge and culture was a necessary accompaniment to conquest. Alexander took with him doctors, zoologists, philosophers and map-makers, and thorough records of their experiences were kept. Above all Alexander established the city which would dominate intellectual life for the next five hundred years. From Alexandria would emerge the classical world's final assessment of the Earth and the universe. For Strabo, a wealthy Greek traveller from the Black Sea, the civilization and erudition of this city on the coast of Egypt were like a magnet. He visited it a few years before the birth of Christ, and his description of it in his seventeen-volume *Geography*, to which we owe so much of our knowledge about Greek views of the world, underlines Alexander's prescience in choosing such a spot to nourish the flowering of culture and academic advance.

Sheltered from the Mediterranean by the island of Pharos and with a deep natural harbor and two promontories into the open sea, Alexandria selected itself as a strategic and mercantile headquarters. Unlike other Egyptian cities, which stifled and stank in the summer sun as the lakes evaporated into mud and marsh, Alexandria, washed on two sides by the sea and cooled by the rising of the Nile and the Etesian winds from the north, remained temperate and welcoming. The city was laid out in two grand avenues intersected by streets made for horse riding and chariot driving. Each king of Alexandria felt obliged to build a new palace for himself so that the royal quarter was a checkerboard of monumental splendor. For hundreds of years it would be the most important conduit for the exotic produce of the East—silk, spices and precious stones. All races and colors mingled in this sophisticated entrepôt. The most important institution for posterity was the Museum, in which

An Egyptian statue of Alexander the Great, whose conquests greatly increased the Greeks' knowledge of the world. His tutor, Aristotle, told Alexander that he could be master of the entire world stretching from the Pillars of Hercules (the Strait of Gibraltar) in the west to India in the east. Alexander saw India, but was thwarted in his ambition to find the limits of Asia.

A camel train crossing the Egyptian desert. The measurement of distance in ancient times was fraught with difficulties. Mapmakers had to rely on the duration of journeys by camel or on foot across land, and by ship over the sea. But nobody could be sure to maintain a steady speed or stay on a straight line.

was housed the Royal Library. This comprehensive think-tank was probably founded by King Ptolemy Soter at the end of the fourth century BC and rapidly expanded under succeeding royal patrons. Little is known about its structure except that it had a public walk where "philosophers, rhetoricians and all others who take delight in studies can engage in disputation," an *exedra* (a raised, open-air platform) with seats, and a large house which contained the common mess-hall of the men of learning who shared the Museum. The library collected books, or "scrolls," from all over the known world covering every aspect of learning—medicine, philosophy, astronomy, and mathematics, as well as literature and poetry. It contained 400,000 multi-volume works and 90,000 single books.

If our knowledge of the building is limited, the brilliance of the minds who gathered there has shone through the ages. In the realm of geography and mapmaking the brightest beacon was Eratosthenes. Born in Cyrene, a few hundred miles west along the coast of North Africa, and later a resident of Athens, he was headhunted by King Ptolemy Euergetes around 235 BC to act as librarian and as tutor to his son. Eratosthenes' writings were lost but it is known from Strabo, who based much of his own work on that of the Cyrenian, albeit with emendations which were often retrograde, that Eratosthenes compiled a comprehensive geography of the known world. However, his fame rests on another achievement—the first scientifically based measurement of the Earth. Until then rough calculations had been made by locating two positions on the same meridian (an imaginary curve running around the Earth through the North and South Poles) at which two easily identifiable stars were at their zenith. Two such stars were the head of the constellation Draco, which sat above the town of Lysimacheia, and a star in Cancer, which was in the zenith over the southern Egyptian town of Syene (now Aswan). The distance between the two stars was reckoned to be one fifteenth of the heavenly sphere. The distance between the two towns was estimated at 20,000 stades, just under 2000 miles (3220 km). Multiply 2000 by 15 and you get 30,000 miles (48,280 km). Not bad, but an overestimate of some 20 per cent.

Eratosthenes had a different idea. He adapted the technique so that the key was the angle at which the sun shone down on a particular point of the Earth. It was reliably reported that on the summer solstice, 21 June, the sun stood directly over Syene. Indeed there was a well in which at midday the sun cast no trace of a shadow. (Today there is an ancient well at Aswan where this part of Eratosthenes' method can be tested. The midday sun does indeed shine more or less directly overhead.) Eratosthenes believed that Alexandria lay directly north of Syene on the same meridian. A camel caravan travelling at around 100 stades (10 miles/16 km) a day needed fifty days to make the journey between the two towns, a distance therefore of 5000 stades (500 miles/800 km). The final piece of the jigsaw was the angle at which the sun shone down in Alexandria at midday. With a standard gnomon this was easy for Eratosthenes to measure. The angle turned out to be just over 7°, about a fiftieth of a circle. Multiply the distance between Alexandria and Syene, 5000 stades, by 50 and you have the Earth's circumference, 250,000 stades (25,000 miles/40,230 km). With the accustomed Greek mania for numerical and symmetrical elegance Eratosthenes rounded this up to 252,000 stades (25,200 miles/40,550 km) so that it could be divided by 60 and make a neat addition to the Earth's known dimensions. Sadly for posterity there is a long-running dispute about the precise length of a Greek stade. A likely figure is just under ten stades to a mile, which gave Eratosthenes a final result of 24,663 miles (39,690 km) compared with the true figure of 24,862 miles (40,010 km). It was an exceptional achievement, arrived at with a few clues, a simple instrument and immense ingenuity.

Eratosthenes was just as interested in the age-old quest to establish the shape of the inhabited world. A few more clues were emerging, particularly to the north. The Phoenicians had remained secretive about their routes to northern Europe because they did not want any rivals for the Scilly Isles and Cornish tin trade. But around the same time as Alexander the Great was pushing into India, a certain Pytheas, from Marseilles, swept aside the freezing curtain to the north. Pytheas aroused extreme mistrust. Strabo went so far as to accuse him of being an ''arch-falsifier.'' Such antagonism was engendered by the outlandish descriptions with which Pytheas returned from his travels. He claimed to have visited Britain and travelled over a considerable part of it. He stated that its coastline was 40,000 stades (4000 miles/6440 km) in length and that the island stretched to the north. The size was an exaggeration but the location was more accurate than Strabo's assertion that Britain lay to the northeast along the coast of France.

Pytheas also claimed that north of Britain he had found another island, which he referred to as Thule. Beyond it the sea assumed a sludgy consistency like that of the jellyfish, or sea-lung. According to Pytheas, this gelatinous medium held the Earth, the sea, and all the elements in suspension, forming a bond on which you could neither walk nor sail. Pytheas reported that on Thule ''the barbarians showed us where the sun goes to his rest, for it happened about these parts that the night was

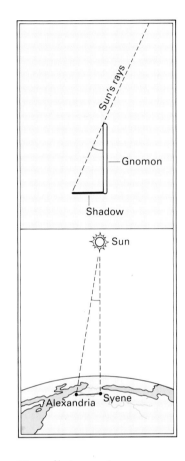

The earliest accurate measurement of the Earth's circumference was made by Eratosthenes in the third century BC. There were three elements in his calculation: the overhead position of the sun at Syene on the summer solstice, the distance between Syene and Alexandria calculated by a camel's progress and the angle of the sun's rays in Alexandria at the same time and date. The result was within 200 miles (320 km) of the figure we know it to be today.

only a little interval after the setting of the sun before it rose again." The intriguing conclusion could be that Pytheas did indeed reach as far north as Iceland or northern Norway during the summer solstice. If so his "sea-lung" was no invention but an accurate description of the sludge which forms along the edge of drift ice at the Arctic Circle. Despite Strabo's attacks, Pytheas' Thule, which may later have been relocated and designated as Iceland, crept into geographic terminology as the northern limit of the inhabited world. All the while the road which began with Hephaestus' shield and the Milesian philosophers was leading in the years before the birth of Christ towards consensus.

Strabo's encyclopedic *Geography*, a cornucopia of the human, animal and physical world written around this time, portrayed this emerging world in unparalleled detail. The Earth was a sphere whose size had been measured with considerable accuracy. It was divided into climatic zones two of which supported life. In one quarter of the northern hemisphere lay the inhabited world whose shape and precise contents were the focus of Strabo's research. One hundred years earlier the Stoic philosopher Crates, who was in charge of a new library at Pergamum which briefly rivalled that at Alexandria, constructed a globe which enabled him to speculate about what might lie elsewhere on Earth. Crates believed that the other three quarters of the world were probably inhabited. For Strabo such intellectual games were intriguing but irrelevant. These areas were unreachable either because they lay across the torrid zone or on the other side of seas so large that they could never be crossed. Today this seems a very narrow outlook. But most parts of the Earth were no more within the reach of the ancient Greeks than are worlds that are light years away from us. Strabo's inhabited world was twice as long as it was broad; a parallelogram which Strabo compared with the cloak known as a *chlamys*, familiar to all his readers. The "Aethiopian" capital of Meroe (Khartoum) and the "cinnamon country" were the southern limits of human life and, because of his mistrust of Pytheas' report of Thule, Strabo defined the northern limit as the island of Ireland, which he located well to the north of Britain. The western border was the Pillars of Hercules, the eastern the sea into which the Ganges flowed.

The more distant the land, the more geographers indulged their imagination. For example, India had been variously described as peopled by "men that sleep in their ears," "men without mouths" and "men without noses." Strabo was critical of such fabrications and unimpressed by stories of the battles between the cranes and the pygmies or accounts of "snakes that swallow oxen and stags, horns and all." Ignorance was the handmaid of invention, and Strabo sought to dispel it. But despite his confident description based on his own experiences and travellers' tales, his map was more cramped than he could ever imagine.

Left *Rackwick Bay in the Orkney islands, off Scotland, on the northern fringes of the world known to the Greeks. The explorer Pytheas claimed to have discovered an island called Thule, north of Britain, where the sun hardly set in the middle of summer – a phenomenon well known to the people of these northerly islands.*

2

IMPERIAL VISIONS

I n the reign of the Emperor Augustus, visitors to Rome were greeted by a map of the known world boldly displayed in the portico of Octavia, his sister, on the Via Flaminia. It had been drawn by Augustus' son-in-law Agrippa and instigated by his adoptive father Julius Caesar, who had ordered that the whole world be measured, partly so that it could be taxed. No one who saw the map could fail to understand that the world was Roman and that its boundaries enclosed the one true civilization. Yet such an observer was being thoroughly deceived.

Augustus advised his successors to confine their domains within the limits which nature seemed to have ordained. To the west was the Atlantic Ocean; to the north the Rhine and the Danube; to the east the Euphrates; and to the south the stifling, inhospitable deserts of Africa and Arabia. There seemed no point in allowing the ordered march of civilization to be disrupted and polluted by the barbarism that lay beyond these borders. Why bother to expand into the forests and morasses of northern Europe or the sandy aridity of Africa, both peopled by primitive tribes, when the sophisticated life radiating out from Rome could be enjoyed in agreeable weather and among friendly, or at least subdued, populations who provided plenty of good cooks and slaves?

Many thousands of miles to the east, undreamed of and untouched by the classical world, another great empire was taking a similar view of its own importance. It too was bound by geography. To the south and east was the sea; to the north lay the Gobi Desert, its fringes bristling with aggressive foraging nomads against whom the only defence was a great wall stretching from the Yellow Sea thousands of miles inland; to the west was the roof of the world, the plateau of Tibet, with the Himalayas receding into the distance, impassable, protective and isolating. China, the "Middle Kingdom," lay proudly in its introverted splendor. In the sixteenth century the Italian Jesuit priest Matteo Ricci, who was to revolutionize the Chinese view of the world, wrote in amazement: "Their universe was limited to their own fifteen provinces, and in the sea painted around it they had placed a few islands to which they gave the names of different kingdoms they had heard of. All of these islands put together would not be as large as the smallest of the Chinese

Left *The Forum, at the heart of Imperial Rome. Visitors to the city in the first century* AD *could see a map of the known world which suggested that everything was ruled by Rome. The empire's boundaries were the Atlantic Ocean to the west; the Rhine and Danube rivers to the north; the Euphrates to the east; and the African desert to the south.*

provinces. With such a limited knowledge, it is evident why they boasted of their kingdom as being the whole world, and why they call it Thienhia, meaning, 'everything under the heavens.' " Was their outlook so very different from that of the Roman world map on the Via Flaminia?

Between them, classical theory and Chinese inventiveness laid the foundations for modern maps of the world. As the best minds of these two great cultures bore down on the puzzles of the world, what were their final achievements and how did they compare? In the two centuries following Augustus and the birth of Christ, classical geography found its summation in the *magna opera* of two sharply contrasting authors. These works sowed the seeds of the conflict which would overtake Christendom for nearly fifteen hundred years, as the fantasies of popular storytellers overshadowed the observations of scientists. Neither writer would have been the easiest dinner guest. A more entertaining although possibly interminable evening might have been spent with the first, Gaius Plinius Secundus, commonly known as Pliny (the Elder).

Born in AD 23 in northern Italy, Pliny became a distinguished public servant and friend of emperors, holding the job of procurator in Spain in the reign of Nero. He was then appointed by the Emperor Vespasian to the command of the Roman fleet stationed in the Bay of Naples. Between his public duties and cultivation of the imperial family, Pliny was an immensely prolific author. He wrote a twenty-volume *History of the Wars in Germany* and a thirty-one volume history of his own times, both of which failed to survive. Pliny also found time for his *Natural History* in thirty-seven volumes. He proudly relates that he collected twenty thousand facts from two thousand volumes which he deemed significant enough to pass on to his readers in his *Natural History*. He boasted that this Herculean research had been carried out in his spare time, mainly at night. Pliny did not significantly depart from the major tenets of the world view so painstakingly developed by the Greek scholars. He encapsulated the apparent gulf between common sense and scientific deduction thus: "Here there is a mighty battle between learning on one side and the common herd on the other: the theory being that human beings are distributed all round the Earth and stand with their feet pointing to each other, while ordinary people enquire

The city of Alexandria, founded in the fourth century BC by Alexander the Great, became the center of Greek science. Eratosthenes, who measured the world's circumference, worked at the library here, as did Ptolemy. Lying at the southeast corner of the Mediterranean, it was a crossroads of ancient civilizations, an ideal focus of information and theories about the world.

why the persons on the opposite side don't fall off. Yet it is surprising that with this vast level expanse of sea and plains the resulting formation is a globe.'' His arguments for the curvature of the oceans' surface were particularly endearing, the most prominent being that hanging drops of liquid always take on the shape of small round globes.

His detailed geography was a dry and interminable catalogue of the names of cities, tribes, rivers and mountains. Such lists might have bored his readers and dinner companions, but they would have perked up when he began to trot out his marvellous tales of strange places, peoples and animals. In the woods and groves beyond the north wind lived the Hyperboreans to whom death came only when, having had enough of life, they held a banquet, anointed their old age with luxury and leapt from a certain rock into the sea. There were the All-ears Islands whose inhabitants had very large ears covering the whole of their bodies, which were otherwise naked. In the middle of the African desert could be found the headless Blemmyae, whose mouth and eyes were on their chest. Along with hundreds more amazing peoples were animals such as the griffin, a wild, winged beast which mined gold; the tailed apes which had been known to play at draughts; the nereids, whose bodies bristled with hair even in the parts which were of human shape; the turtles of the Indian Ocean, which could be used as a roof for a house one way up and as a sailing boat the other. Pliny stimulated an enduring interest in fabulous geography and in the following centuries this view of the world took popular precedence over scientific and descriptive exactitude. The contents of Pliny's world would be scattered over maps, often in beautiful illustrations, for over sixteen hundred years.

Pliny's opposite was the most important scientific accumulator of his time, Claudius Ptolemy, one of antiquity's most shadowy figures. For a man whose influence was, over a thousand years later, to be so massive, the records of his life are frustratingly sparse. He was born in Egypt and lived and wrote at Alexandria in the middle of the second century AD. He wrote two great works, the *Almagest*, which was a treatise on astronomy, and the *Geography*. He would probably have sent his dinner companions to sleep, for unlike Pliny, he had no interest in amusing tales or colorful description. Ptolemy was a copyist on a gigantic scale, although he had the grace to acknowledge generously his debts to his predecessors. In astronomy, the most important of these was Hipparchus, who had worked in Rhodes in the middle of the second century BC, about three hundred years earlier. Hipparchus catalogued at least 850 stars and laid the foundations for a precise map of the heavens. It is to Ptolemy's credit that the work of this unrivalled stargazer has reached us down the ages.

Hipparchus' detailed observations had put further strain on the creaking cosmology of Earth-centeredness. As Ptolemy set to work on his own development of Hipparchus' data he too was blinded by geocentricity. ''If one considers the position of the Earth,'' he wrote in the *Almagest*,''one will find that the phenomena associated with it could take place only if we assume that it is in the middle of the heavens, like the center of the sphere.'' The core of his argument rested on the symmetry of daylight patterns over a year. If the Earth was not at the center the equinox would not occur at equal intervals between the summer and winter solstices. We must be in the middle because the universe above and below us, or to the front and back, has to be divided into equal parts. Ptolemy was not merely interested in copying and recording heavenly phenomena; he also wanted to become the perfect system builder. As Alexander wished to rule the whole inhabited world, Ptolemy had dreams of presiding over the completed model of the working universe. But improved

The Greek astronomer Hipparchus taking observations from his observatory in Alexandria. Hipparchus catalogued over 850 stars and laid the foundations for a precise map of the heavens. His detailed sightings of the irregular movements of heavenly bodies challenged the Greek idea that everything revolved in perfect spheres. Hipparchus also divided the world into degrees of latitude and longitude.

Ptolemy, the astronomer and geographer whose reputation outweighed all others because, by a historical fluke, his two major works, the Almagest *and the* Geography, *survived. Ptolemy was a copyist on a gigantic scale, whose debts to his predecessors like Hipparchus and Marinus of Tyre were condescendingly acknowledged in his writings.*

observations were causing more and more difficulty. The problem was not simply the erratic wanderings of the planets; it was apparent that the most frequent pattern described by heavenly bodies was an oval or ellipse rather than a circle.

Ptolemy was blinded by the all-powerful belief that the sphere was the perfect shape, and this ideology took precedence over observation in his work. Even Ptolemy, who professed himself so anxious to work from experience rather than imagination, was forced to yield to tradition. The result was a yet more elaborate concoction of spheres—in its way an edifice of genius, but flawed at its very heart. In addition to the basic cycles described by the sun, moon, planets and fixed stars around the Earth, three times as many epicycles as before were incorporated into the system. Arthur Koestler compares this assembly with a Ferris wheel at an amusement park. The basic circle is the movement of the great wheel. Then each chair itself starts revolving. "The unhappy passenger—or planet—is now describing a curve in space which is not a circle, but is nevertheless produced by a combination

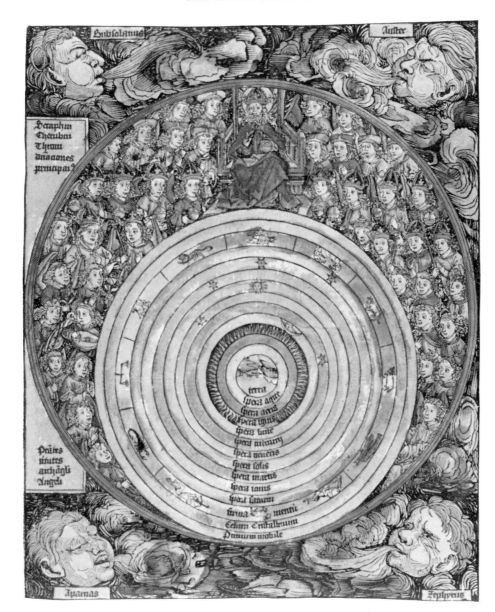

An engraving dated 1493 shows the Ptolemaic system of the universe. Earth is at the center, with the moon, sun, planets and stars revolving around it. Their exact movements and positions were catalogued in Ptolemy's book the Almagest, *a complex compendium of scientific and astronomical calculations. This work demonstrated Ptolemy's genius as a mathematician and geometer.*

of circular motions. By varying the size of the wheel, the length of the arm by which the cabin is suspended, and the speeds of the two rotations, an amazing variety of curves can be produced, kidney-shaped curves, garlands, ovals and even straight lines.'' It was ingenious but crazy. Ptolemy's universe could be mapped on star charts, or represented in celestial globes and armillary spheres (used to track the movements of planets).

The Earth's position in the universe having been decided, what then lay on it? Here too Ptolemy was indebted to Hipparchus, who had conceived the simple but revolutionary innovation of dividing the length and breadth of the world into 360 degrees. Within this framework Hipparchus decreed that every important place on Earth could be fixed by the point at which the east-west curve of latitude and the north-south curve of longitude intersected, as accurately measured by astronomical observation. This perfectionist approach was unfortunately subject to the constraints of practicality. It was difficult for travellers to remember their gnomons, and

An armillary sphere, showing the supposed revolutions of the heavenly bodies around the Earth. These devices were immensely popular in the Renaissance, but the picture they created was extraordinarily intricate because of Ptolemy's insistence that everything moved in spheres. To account for the zigzag and elliptical paths observed by astronomers, a whole series of epicycles and spheres within spheres had to be created.

take measurements of the sun in far-flung corners of the world. By Ptolemy's time there was an additional tool, the astrolabe. This was a round piece of flat metal or wood whose perimeter was divided into 360 degrees. From its center there rotated a straight arm, which was lined up against a star, allowing the angle of elevation to be read off. The astrolabe was more accurate and flexible than the gnomon, as it did not depend on the casting of a shadow and could be used to record the features of the night sky.

Hardly a part of everyday baggage, the astrolabe did nothing to overcome the obstacle of longitude. Angular measurements of the sun and stars fixed latitude but were of no use for determining longitude. The longitude of a given point is expressed as the angular distance east or west of a central or prime meridian. Today the prime meridian is an imaginary curve drawn from the North Pole through the Royal Observatory at Greenwich in London down to the South Pole. So Greenwich's longitude is 0° and New York's, for example, is just under 74° west. In Ptolemy's era the prime meridian went through the Fortunate Islands (the Canaries) which had replaced the Pillars of Hercules as the western edge of the inhabited world.

The Greeks understood that the key to longitude was time. The sun's daily rotation around the Earth (in reality the daily rotation of the Earth on its axis) meant that, from any given point, daybreak occurred earlier the further east you went and later the further west. For example, Delhi is five and a half hours ahead of London and New York five hours behind. Therefore degrees of longitude could also be expressed as time. Twenty-four hours divided by 360 equals four minutes, which equals one degree of longitude. Finer accuracies can be expressed in minutes and seconds of longitude; the precise longitude of New York is 73° 59′ 31″ west.

However, the only way to establish longitude is to compare local time with the time at the prime meridian, a hindrance which Hipparchus fully acknowledged. He realized that the solution lay in taking the time in different parts of the world during a single event visible everywhere—an event such as a lunar or solar eclipse. But how on Earth, particularly the classical Earth of slow communications, could you arrange to have people scattered among towns, cities and bends in rivers checking the time (a crude operation without mechanical clocks) during infrequent eclipses? It was brilliant in theory but totally unworkable. Instead distances had to be measured by the number of days the journey took on foot, or by horse, camel or boat. It was inevitably imprecise, for itineraries do not necessarily follow straight lines and neither humans nor their animals are metronomic machines. Nevertheless Ptolemy, in addition to the building of his universal system, set himself the task of creating the perfect world map, and thereby left a trail of confusion down the ages.

Ptolemy's first error concerned the size of the world. Some of Eratosthenes' successors decided that they knew better than the master himself, but such arrogance was their undoing. The chief troublemaker was Posidonius, a Stoic philosopher who was born in Syria and settled in Rhodes. He reverted to the old method of comparing the elevation of a particular star at two places on the same meridian. The star was Canopus, which was only just visible on the horizon at Rhodes and deemed to be one forty-eighth of a circle higher when seen from Alexandria. The latter city was assumed, wrongly, to be on the same meridian and, also wrongly, to be 5000 stades (500 miles/800 km) from Rhodes. Five thousand times forty-eight equals 240,000 stades (24,000 miles/38,620 km), slightly less than Eratosthenes' figure but not too bad. Even this degree of accuracy was achieved only because the inaccuracies in all parts of the calculation cancelled each other out. Posidonius then sought to correct some of his errors, notably by reducing the distance from Rhodes to Alexandria to 3750 stades (375 miles/600 km), but letting other mistakes stand. His final figure was 180,000 stades (18,000 miles/28,970 km), a massive reduction on that of Eratosthenes. Posidonius' methods were inferior, his conclusion retrograde. Yet it was the one adopted by Ptolemy, and this belief in a smaller world was the impetus for Columbus' voyage sixteen hundred years later.

Ptolemy, sitting in the candle-lit corners of Alexandria's great library, hunched over his quill and parchment rolls, a pernickety man who copied his predecessors' calculations while delighting in pointing out the minutiae of their errors, pursued his overweening ambition of creating the total world map. Be sceptical, distrust travellers' tales, beware "their love of boasting" which makes them magnify distances, warned Ptolemy. This scepticism was all a front. Ptolemy filled six large volumes with the coordinates of latitude and longitude of just about everywhere known on earth. He must have known how inaccurate much of this information was and yet he set it down unquestioningly. The process was easier than it might appear. Ptolemy had copied much of the information from Marinus of Tyre, a geographer who compiled it a few decades before and whose work did not survive. Ptolemy expiated his debt by crediting his source but left the reader in no doubt about his own superiority. But Ptolemy's intellectual vanity backfired, for part of his legacy was the perpetuation of Marinus' mistakes.

Apart from the unnecessary adoption of a reduced world, these mistakes were in general understandable. The inhabited world was apparently a larger area than previously thought. Despite the injunction of Augustus, the Emperor Trajan had spread by military conquest the tentacles of the Roman Empire. First he crossed the Danube to crush the Dacians, renowned for their thirst for battle. After five years of war and thousands of Roman deaths, a new province, 1300 miles (2090 km) in circumference, was added to the Empire. Trajan next headed east. He defeated the Parthians and then sailed in triumph down the Tigris to become the first and last Roman general to navigate the Persian Gulf. Armenia, Mesopotamia and Assyria were reduced to the status of Roman provinces. From these distant outposts trade routes extended frontiers even further. There seemed to be never-ending land beyond India to the east. The same was true of Africa. Ptolemy wrote that one Roman general had marched for four months south of the cinnamon country, the traditional end of the inhabited world, and arrived at a country inhabited by Ethiopians, called Agisymba, in which rhinoceroses abounded. The expedition must have struck deep into the heart of Africa. On such evidence Marinus had substantially increased the length and breadth of the inhabited world.

Ptolemy reined back these dimensions but kept intact the general composition of Marinus' redefined world. The key change was to banish the age-old idea that Africa was surrounded by the Ocean Sea. Instead it was extended without limit to the south and its eastern coast curled round to stretch thousands of miles east, eventually linking up with Asia. The Indian Ocean was enclosed, a giant lake in the center of the world landmass. To the south was Ptolemy's famous *Terra Australis Incognita*, the unknown land which would excite so many fantasies and attract so many explorers more than a millennium and a half later. The idea had been aroused by envisaging the known world on a globe. Without a corresponding landmass on the other side the Earth would have been unacceptably lopsided, an affront to the classical love of symmetry. A hundred years earlier it had been named the Antichthon, the counter-earth, by the geographer Pomponius Mela.

Ptolemy's innovation was the representation of this world on a map. Globes, being the same shape as Earth, presented few problems. You simply built a ball,

A battle scene on the Emperor Trajan's column in Rome. Trajan extended the Roman Empire by military conquest, bringing new knowledge of remote parts. He even dared to cross the Danube and fight the Dacians, renowned for their thirst for battle. The English historian Gibbon described the Dacians thus: "To the strength and fierceness of barbarians they added a contempt for life, which was derived from a warm persuasion of the immortality and transmigration of the soul."

The world of Ptolemy recreated in the edition of his Geography *printed in Ulm, southern Germany, in 1486. Ptolemy's most important innovation was to banish the age-old idea that Africa was surrounded by an ocean. Instead he extended it to the south and made the Indian Ocean an enormous landlocked lake. This was an obstacle for European sailors of the fifteenth century searching for a passage round Africa to the east.*

drew the lines of latitude and longitude around it and their intersections gave precise coordinates for any part of the world. By Ptolemy's time the inhabited world was becoming too large and knowledge of its contents too detailed to allow easy display on the restricted size of a globe. By contrast, maps could be divided into continents, regions, even towns, and drawn to large or small scale. But how could the curved meridians and parallels be projected onto a flat piece of parchment? Until Ptolemy it had been done simply with a grid of straight lines intersecting at right angles. This would have worked perfectly for a flat Earth, even for a small area of our round Earth where the curvature of its surface was insignificant. Used over the whole world, however, the grid led to extreme distortions. The solution was to draw some lines as curves. Ptolemy's first projection had straight meridians fanning out from the North Pole and intersecting with parallels drawn as arcs of circles. His second projection took the idea one stage further—meridians and parallels were both curved to produce the sensation of looking at a fully rounded world. Theoretically it was an improvement, but it rendered the task of drawing the map more difficult. North-south distances could no longer be plotted with a rule, so Ptolemy kept the first projection in reserve "for the sake of those who will have recourse to the earlier projection because of indolence."

Thirteen hundred years later these projections and Ptolemy's coordinates would take the intellectual world by storm as they resurfaced in exquisitely hand-drawn Renaissance reconstructions. His great achievements were to provide a fully scientific framework and to popularize the grid of latitude and longitude from which maps are made today. His astronomic phraseology still survives when we talk of the rising and setting of the sun. His star charts are the foundation for modern navigation. But, as E. H. Bunbury concludes in his *Ancient Geography*, Ptolemy's science was a mirage. "He saw clearly the true principles upon which geography should be based, and the true mode in which a map should be constructed. But the means at his command did not enable him to carry his ideas into execution; the substance did not correspond to the form; and the specious edifice that he reared served by its external symmetry to conceal the imperfect character of its foundations and the rottenness of its materials."

While Ptolemy battled with his calculations, he had no inkling that the scholars of another great culture thousands of miles away were having more success in mapping their lands. Pliny, however, had hinted at the existence of the land of the Seres, the silk people, lying at the eastern edge of the inhabited world. The link with them was the enchanting island of Taprobane (Sri Lanka, formerly Ceylon), whose size was hugely exaggerated by the ancients. According to Pliny, the inhabitants of Taprobane lived in a state of Utopian happiness and led so healthy a life that they often reached the age of a hundred. Gold and silver, pearls and precious stones abounded in this idyllic place. From there an envoy called Rachias was supposed, while on a trading mission, to have visited the Seres whom he described as men of gigantic stature with red hair and blue eyes. Rachias was mistaken, for such a description hardly applied to the Chinese. In an anonymous periplus, a description of seas and coastlines, written shortly after Pliny's death in 79 AD, more clues appeared. The periplus listed the coastlines of Arabia and India right round to the mouth of the Ganges. Opposite lay an island called Chryse, which produced the finest tortoiseshell in any part of the Indian Ocean. Then came the crucial information: "Beyond this country, lying quite up to the North, where the seacoast ends externally at a place in the region of Thina, was a city in the interior called Thinae, of very great size, from which was exported silk, both in the raw state and spun, and woven into fine stuffs: these were carried to Barygaza overland through Bactria, and on the other hand down the river Ganges to Limyrice."

The periplus was a landmark, the first mention in classical documents of China and the first description of the silk route. Silk itself was well known and highly valued in the West. The Chinese tell a story of its first appearance in Rome. Julius Caesar went to the theater wearing an imperial coat woven from Chinese silk. The audience was so entranced by its beauty that they could not take their eyes off the emperor's magnificent apparel and saw nothing of the play. They had no inkling of the origin of the silk, however. The world view of the Greek and Roman civilizations was the product of their vantage point in the Mediterranean. In particular, it resulted from the imagination, guesswork, and scientific deduction of individuals in cities dotted around its shores. The Chinese vantage point was very different. In China two thousand years ago the mysteries of the heavens would have exerted just the same attraction as they did on the Greeks. But for the Chinese astronomer life was very different. By contrast with the individualism of the Greeks, Chinese science was institutionalized. There were no independent schools of philosophy like those of Pythagoras in Kroton or Aristotle in Athens. By contrast, the Chinese astronomers worked for the government.

In the period of the "warring states," a few centuries BC, Confucius had wandered throughout the fragmented states of China advising their rulers on the right way to run a country. Princes were exhorted to rule with benevolence and sincerity, eschewing force and distributing social justice. Form, propriety, tradition and respect for one's ancestors were the keys to harmonious rule. Confucius founded a school where an elite corps of public servants could be trained to help rulers govern by these principles. Gradually his ideas took hold and bureaucracies developed first in individual states and then, as China became unified under the Han Dynasty, at the center of empire.

Bamiyan in Afghanistan, on the Silk Road from China to the West. For centuries, trade in silk linked the two great civilizations of China and the Mediterranean basin. A Chinese story tells how when Julius Caesar visited the theater in a coat woven from Chinese silk the audience was so entranced by its beauty that they could not take their eyes off his apparel, and ignored the play.

Astronomical calculations were a vital imperial tool. They were not only the foundation of knowledge about the Earth and the universe, but also the source of the calendar. The latter's acceptance by the agricultural people of China amounted to an acknowledgement of allegiance to the emperor and new emperors reordered the calendar to exert their authority. While Western civilizations ebbed and flowed, Chinese astronomers kept the longest continuous astronomical records of any society. From the fifth century BC to the tenth century AD Chinese records are almost the only ones available. Joseph Needham, in his monumental *Science and Civilization in China*, to which this account of Chinese progress is indebted, states that the Chinese alone recorded the appearances of novae and supernovae for the whole period between Hipparchus and the late-sixteenth-century Danish astronomer Tycho Brahe, when no one else knew or wished to know that "new stars" sometimes appeared in the heavens.

State administrations were keen to collect the best minds around them. In the third century BC the lord of Ch'in was ashamed that his state was not equal to others in scholarship. He enticed scholars to come, offering them lavish entertainment. Eventually three thousand visitors turned up. These guests were asked to file their observations, which were published in the "Spring and Autumn Annals" and displayed at the gate of the state capital's marketplace. Alongside were "1000 catties of gold" with a notice informing travellers and visitors that anyone among them who could add or subtract a word would win the treasure. Occasionally indolence and dissimulation interfered with the smooth flow of record. In the eleventh century AD the Astronomer Royal found that his two observers were not bothering to make any observations but simply copied out each other's reports for previous years. And one of his successors found little to admire in the examination procedure. "The Ministry of Rites arranged for the examination candidates to be asked to write essays on the instruments used for gaining knowledge of the heavens. But the scholars could only write confusedly about the celestial globe. However, as the examiners themselves knew nothing about the subject either, they passed them all with a high class."

The Chinese possessed a wealth of astronomical knowledge. But what conclusions did they come to about the world and the universe? Were they ahead of or behind the Greeks? It is difficult to determine who won this scientific race as the Chinese developed three distinct cosmological ideas, all of them emerging from the flourishing intellectual and speculative climate in the final centuries before Christ, the period between Pythagoras and Ptolemy. Despite their isolation, some of these ideas bear tantalizing resemblances to those of Western civilizations. The oldest theory was the Kai Thien in which the heavenly vault was hemispherical in shape and the Earth was an inverted bowl. Joseph Needham states that this double-vault theory may well have been borrowed from Babylonian cosmology. As the Chinese astronomers later used Babylonian star observations, the idea of such a transmission is enticing. Elsewhere in the Chinese chronicles of the Kai Thien the heavens were round in shape like an open umbrella while the Earth was square like a chess board. The constellation of the Great Bear was in the middle of the heavens and the inhabited world in the middle of the Earth. Rain falling on the Earth flowed down to its four edges from which it fell off to form the surrounding ocean. The sun, as it moved around, lit up different parts of the Earth like a moving spotlight.

The second theory, the Hun Thien or Celestial Sphere, invites immediate comparison with Greek ideas. It was known to be current in China at the end of the second century BC, a few hundred years after Pythagoras had propounded the spherical movement of the heavens. The man who first wrote about it, at the

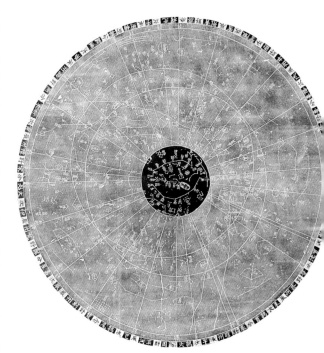

beginning of the second century AD, was one of China's most remarkable polymaths, Chang Heng. He invented a seismograph which could detect earth tremors many hundred of miles away and he improved the design of armillary spheres. He even wrote poetry. But it is as an astronomer and mapmaker that he influences this story. "The heavens are like a hen's egg and as round as a crossbow bullet," wrote Chang Heng. "Heaven is large and the Earth small. Inside the lower part of the heavens there is water. The heavens are supported by vapour, the Earth floats on the waters." The similarities with the Western world view are uncanny. The Egyptians and the Babylonians had both seen the world as an oyster surrounded by water; the egg also had been used as a symbol in Egypt. Thales' world rested on water until it had been released into space by his successors. Around this Chinese Earth the heavens revolved in their circular orbits, an idea which came to astronomers all over the world from simple observation of the skies.

The most exciting idea of all was the Hsuan Yeh or "infinite empty space." The masters of this school had taught that "the heavens were empty and void of substance. When we look up at them we can see that they are immensely high and far away, having no bounds. The human eye is colorblind and the pupil near-sighted; this is why the heavens appear deeply blue. The sun, moon, and the company of the stars float freely in the empty space, moving or standing still." If such a theory had been propounded in the West at any time in the nearly two and a half thousand years between Homer and Kepler it would have caused intellectual chaos. Infinity and motionless stars and sun cut ruthlessly through Ptolemy's cycles and epicycles.

Such ideas were complemented by those of the Buddhist scholars who were spreading into China from India. They wrote that in the whole of empty space heaven and Earth were just a small grain of rice. They even hinted at human life elsewhere in the great expanse of the universe. "How unreasonable it would be," said one, "to suppose that besides the heaven and Earth which we can see there are no other heavens and no other earths!" Chinese documents do not record whether or not they ever received visitors from outer space. Despite these theories and the sophisticated astronomical calculations which accompanied them the Earth stayed stubbornly flat and square. Some scholars challenged this orthodoxy on the grounds that it was difficult to see how its four corners could be fully covered. The movements of the heavens would be hindered by this square-edged and pointed shape jutting unattractively into space. However, these ideas never intruded into the Chinese vision of their world and empire. Whereas the Greeks could draw on the Egyptians and Babylonians, and the Romans on the Greeks, the Chinese had only their own ancient civilization stretching back two thousand years. There was no evidence of cultural centers beyond the seas, deserts and mountains and no reason to suspect their existence. The imperial court, secure in its belief in its own superiority, deliberately restricted its outlook. Its requirement was secure frontiers; border campaigns only served to harden the conviction that the further you went from the Middle Kingdom the more barbaric people became. Life was comfortable. Iron-casting, which was practised in China some two thousand years before it was in the West, enabled all sorts of sophisticated tools to be made for farming—sickles, spades, hoes and ploughs among many others. Agriculture was so efficient that, unlike in Greece and Rome, slaves were not needed in the fields. There was plenty of food, an expanding network of canals for irrigation, an attractive range of silk garments, and plenty of tortoise shells from which to tell omens. The only threats on the horizon were the warlike tribes to the north and east, but the Great Wall was a reassuring deterrent.

Left *A Chinese star chart from 1193. While Western civilizations rose and fell, the Chinese kept the longest continuous astronomical records of any society. One Chinese concept of the universe, called Hsuan Yeh, was remarkably prophetic of modern understanding. It suggested that the universe was infinite empty space in which the sun, moon and stars float freely, moving or standing still.*

Yet there were disadvantages to being the only civilized people on Earth, particularly when it came to marrying off the emperor. In Europe there were always plenty of foreign princesses to choose from in its many fragmented kingdoms. In China the choice of bride always ran the risk of creating internecine rivalries; there were no comparable foreign countries in which to find a consort who did not interfere.

What clues could be found in ancient literature about the shape of the world? The oldest account was the Tribute of Yu, the legendary hero-emperor or Great Engineer who mastered the waters and became the patron of canal engineers and irrigation engineers. "The inundating waters seemed to assail the heavens," said Yu, "and in their extent embraced the hills and overtopped the great mounds, so that the people were bewildered and overwhelmed. I opened passages for the streams throughout the nine provinces and conducted them to the seas." Yu described the nine provinces of China. Yen, for example, had mulberry grounds stocked with silkworms, rich, black soil and luxuriant grass. Beyond these lay the zone of allied barbarians and then the zone of cultureless savagery. The local color was splendid but the broad picture rather hazy.

Help might be sought from the old story of the commission given by the legendary emperor Yao to the astronomers and brothers Hsi and Ho sometime between the eighth and fifth centuries BC. "In reverent accordance with the august heavens," as the commission put it, the brothers were sent off to the four corners of the Earth to delineate the sun, moon, stars and celestial markers, "so to deliver respectfully the seasons to be observed by the people." Hsi was ordered to head east where he was to live among the Yu barbarians and "receive as a guest the rising sun." Then he had to go south and "pay respectful attention to the (summer) solstice." Ho was to travel west to the place called Mei-Ku and "bid farewell respectfully to the rising sun," and then head north to Yu-Tu. This also was a vague picture of the world. In fact the commission, written at the same time as Homer's description of Achilles' shield, is an intriguing allegory. Joseph Needham reveals that Hsi-Ho is not the name of a people but of a mythological being who is sometimes the mother and sometimes the charioteer of the sun. The name was then split up to denote four magicians whose job was to make sure that the sun kept to its regular patterns and so maintain the flow of the seasons, an important matter for an agricultural society.

In the second century BC, under the unified Han Dynasty, the Chinese tried to broaden their horizon. In 138 BC the Han emperor ordered the envoy Chang Ch'ien to go west in search of allies to help fight the Hsiung-nu, the Huns, who would later devastate much of the Roman Empire. According to Chinese records, "At that time the Son of Heaven made enquiries among those Hsiung-nu who had surrendered and they all reported that the Hsiung-nu had overcome the king of the Yueh-chih and made a drinking-vessel out of his skull." Chang Ch'ien's mission was to find the Yueh-chih, who were understandably angry at this peremptory treatment of their king. On his way Chang Ch'ien was arrested by another tribe, the Shanyu, and held for ten years. By the time he reached the Yueh-chih in Bactria (today part of Afghanistan), they had dropped their idea of a war of vengeance against the Hsiung-nu. Chang Ch'ien stayed for a year to look around. He was amazed to find "cities, mansions and houses as in China." He had encountered a remnant of Greek civilization, Bactria having been conquered and settled by Alexander the Great on his way to India. Chang Ch'ien returned with information about Parthia and Syria, perhaps also Egypt and Babylon. Over the next two centuries Chinese armies campaigned in central Asia, picking up often garbled news of the Roman Empire to the west. They called it "Ta Ts'in," meaning that the Romans were as civilized as the Chinese themselves but taller in stature. It was a considerable compliment from the people of the Middle Kingdom. They discovered the Roman demand for silk and tried to set up a direct trade, cutting out the Persian middlemen. The difficulties of the terrain prevented them from finding an alternative route.

Further contact was made when an embassy from the Roman Empire arrived in China in AD 166. The Chinese were so unimpressed with their presents that they viewed the ambassadors as mere merchants. Later, Chinese travellers and monks travelled throughout India and southern Asia and massive and detailed geographies of these regions were written. But the link with the West was never made. The Chinese admired what they knew of the Romans but their introversion was unchanging. In the end they would become the discovered, first by Marco Polo, then by the Portuguese and other Europeans.

If the world picture remained firmly omphalosceptic, Chinese genius asserted itself in masterly fashion over the local mapping of the empire. Silk not only produced pretty clothes—it had a purpose which would have intrigued Ptolemy. In 1973 archaeologists excavated three perfectly intact tombs dating from the Early Han Dynasty in Ch'ang-sha, the capital of the province of Hunan. They contained the body of Li Cang, the prime minister of Ch'ang-sha, his wife and one of their sons. The coffins were decorated with dragons, leopards and mythological animals floating in the clouds with the souls of the immortals. Bamboo baskets replete with meat and fruit and pots of vegetables and cakes, along with recipes for the family's favorite dishes, had been left near the dead to nourish them on the way to the afterlife. Also in the tombs were three silk maps, still decipherable after well over two thousand years. An eyewitness described them as "fragmented like crushed bean curds." The maps were detailed and precise. One showed towns and villages, the second was a garrison map, and the third a relief map. Detailed analysis by the Chinese geographer Kuei-Shang Chang demonstrates that their purpose was strategic and military. For example, they showed rugged terrain and narrow passages vital to the security of Ch'ang-sha, which at the time was on constant alert for invasions from the south.

Chinese familiarity with maps is accidentally confirmed in the story of the assassination attempt on the Emperor Zheng in 227 BC. An agent, Jing Ke, arrived from a neighboring state claiming to possess a map with important military information. The chronicles relate that "King Zheng asked to see the map. Thereupon Jing Ke took out the map, unrolled it and exposed the dagger. Seizing the sleeve of the Qin king with his left hand, Jing Ke grasped the dagger with the right and struck at him. In alarm King Zheng leapt backwards so that his sleeve tore off. Though he tried hard, the king was unable to draw his sword, which was very long. Jing Ke pursued the Qin king, who ran round a pillar. The astounded courtiers were simply paralysed." The courtiers were not allowed to carry arms and the royal guard could not enter the audience chamber unless summoned. But eventually the king managed to draw his sword and wounded his assailant, who was later executed. The story may cast doubt on court security and the efficiency of assassins but it confirms that in China maps were a crucial part of administrative and military planning. For over a thousand years their accuracy was unrivalled anywhere else in the world. The key was the rectangular grid. Stories abound as to its invention, but the innovative Chang Heng is the most likely originator as it was written that he "cast a network [of coordinates] about heaven and earth and reckoned on the basis of it."

More beguiling is the story of Lady Zhao, the younger sister of the prime minister of Wu state. The emperor had often lamented that the nearby states of Wei and Shu

A Chinese map made in the eighteenth century, from a manuscript atlas of Kiangsi province in southeast China. The Chinese mapping tradition began over 2000 years ago, as a form of imperial control. If the emperor knew the exact layout of his land his mandarins could rule it more efficiently.

had not been conquered. To facilitate the movement of troops, he decided that he needed an expert painter to draw a map showing their mountains, rivers and every other physical feature. The young lady, known in the palace as the matchless weaver, was granted the commission. She said: "It is extremely easy for pigments to fade; they cannot be preserved for long. But I can embroider a map." She illustrated the principalities on a square piece of silk, drawing the mountains, the Yellow River, the sea, the cities and the disposition of troops. As she was doing this it struck her that the warp and weft of the silk provided the ideal coordinates for a grid. Whether or not this story is apocryphal, the rectangular grid system was there to stay, enabling accurate scale maps to be drawn. The driving force behind this new and precise science was Phei Hsiu, who lived a hundred years later than his Western counterpart, Ptolemy. In 267 Phei Hsiu was appointed Minister of Works and, in bureaucratic prose which rivalled that of Ptolemy in dryness, he presented his conclusions to the emperor. He found that the origin of maps stretched back hundreds of years but that the Han people had destroyed everything from before their time. On those which had survived from the later Han period his judgement was caustic: "Their arrangement is very rough and imperfect, and one cannot rely on them. Some of them contain absurdities, irrelevancies and exaggerations, which are not in accord with reality, and which should be banished by good sense." He then explained in detail how to use a grid, get the angles and scale right and represent differing heights.

The tradition was established. Large and beautiful maps were created, one being made from eighty rolls of fine silk. Another, the Map of both Chinese and Barbarian Peoples within the Four Seas, measured 30 ft (7 m) long by 33 ft (10 m) high. The Chinese invention of paper encouraged the carving of maps on stone steles from which repeated paper rubbings could be made. The first printed maps appeared in China two hundred years before that technology came to Europe. Gnomons and water levellers, along with astronomical coordinates, produced great accuracy. The Chinese invented the cross-staff, an offshoot of the crossbow, which was their standard weapon in battle long before medieval Europe took it up. It was particularly useful for pinging down arrows from the Great Wall at the northern nomads. One result of these advanced techniques was the Map of the Tracks of Yu, carved in 1137. It is an extraordinarily precise representation of the Chinese coastline and interior rivers. There was simply no equivalent at that time in the West. Ironically their very accuracy may be the reason why some maps from this period have in recent years been "lost." They could be used as evidence that territories, such as Mongolia, are more recent additions to the Chinese empire, now Communist state, than the authorities would like us to believe.

The apex of Chinese ingenuity stemmed from their most momentous discovery, magnetism. The attractive power of the lodestone was known in both East and West, and many legends grew up around it. One was the existence of magnetic islands which ships could not pass if they were built with nails (Ptolemy located these islands between Sri Lanka and Malaysia). The achievement of the Chinese was to harness this phenomenon and invent the compass. The breakthrough was directly related to the Chinese pseudoscience of *feng shui* or geomancy. This was defined as "the art of adapting residences of the living and tombs for the dead so as to cooperate and harmonize with the local currents of the cosmic breath." *Feng shui* is still an integral part of Chinese building design. The spectacular Hong Kong and Shanghai Bank in Hong Kong was laid out in accordance with the dictates of a geomancer. The great bronze lions guarding the entrance had to be set in a strict alignment; in particular they must not look directly towards each other nor be positioned back to back. Inside, wastepaper baskets and desks, chairs and mirrors were adjusted to ward off any evil influence from metal beams above.

As had been the practice for over two thousand years, these calculations for the modern building were made on the geomancer's divining board. Relics from the Han Dynasty show that these boards were made of two plates. The lower one was square to symbolize the flat Chinese Earth and the upper one round to represent the heavenly sphere. This upper plate revolved on a central pivot. Around its edge were engraved the twenty-four compass points. In the center was a representation of the constellation of the Great Plough. The Chinese had noticed that the handle of the Plough, when observed at a constant hour of the night, moved round a complete circuit over the course of the year. It was the oldest pointer of all and its shape in the sky could dictate the way the diviner's board was held.

The leap of the imagination was to put a moving magnetic object at the center of the diviner's board to replace the Plough as the direction-finder. The object was a piece of lodestone cast into the shape of a spoon whose handle and bowl resembled the Plough. The "south-pointing spoon," which is mentioned in Han literature,

Wangshi Gardens in the Chinese silk city of Suzhou. The Chinese map grid is said to have been invented by a young lady who was weaving a military map for the emperor of her state. As she wove, it struck her that the warp and weft of the silk provided the ideal set of coordinates. It is more likely that the innovator was the astronomer Chang Heng, who "cast a network of coordinates about heaven and earth and reckoned on the basis of it".

could then swivel on the diviner's board and the compass points could be correctly set. Joseph Needham investigated how the idea of putting a magnetized moving object on the board might have arisen. His intriguing conclusion is that the origin lay in the ancient war game of chess. The Han records describe an incident in 113 BC in which the emperor asked a magician called Luan Ta to "demonstrate one of his lesser arts by making chessmen fight automatically, and indeed they did mutually hit against each other." What was making them move? A later chronicle records: "The lodestone lifts chessmen. The blood of a cock is rubbed up with needle-iron and pounded to mix. Then when lodestone chessmen are set up on the board, they will move of themselves, and bump against each other." The force was magnetic attraction.

Furthermore there was an ancient form of "image-chess" in which the pieces were astronomic and astrological pieces. The stars then fought each other on the chess board in what Needham described as "a mimicry of the eternal contest between the two great forces in the universe, Yin and Yang." The general balance between these forces in the general cosmic situation could in this way be determined. It is a fascinating argument. Chess pieces represent heavenly bodies. They are magnetized and move automatically on a chess board. Transfer the idea to the illustration of the Plough on the diviner's board and you end up with the lodestone spoon swivelling around to locate the magnetic poles. Spoons gave way to magnetized fishes and turtles and finally to the needle. By around the fifth century AD the needle compass

The waterfront at the Chinese port of Canton, with European ships at anchor, waiting to carry home eastern luxuries. At one time it seemed that Chinese ships could make trading voyages to the West, but the Ming emperors banned such contact with the outside world after the great expeditions of Admiral Cheng Ho in the fifteenth century, and the initiative passed to European traders.

was being used by the Chinese on land. The floating or water-compass followed and then the dry suspension of the needle from below. The Chinese were slow to use the compass at sea, perhaps because the main thrust of their navigation was through inland waterways and canals. But it was being used on ocean-going ships in the eleventh century AD, a hundred years before its arrival in Europe towards the end of the twelfth century.

The outstanding practical result of this Chinese technology was the great expeditions at the beginning of the fifteenth century under the admiral, Cheng Ho, who was known as the Three-Jewel Eunuch. In 1405 sixty-two great ships bearing a vast amount of gold and other treasures set sail from Nanking for the Indian Ocean. No records of these ships remain, but a huge rudder post was found in 1962 near Nanking. It suggests that the rudder alone must have been well over 20 ft (6 m) square. Certainly the ships needed to be enormous for the chronicles record that they carried a force of 37,000 men. Yet this was not primarily a military expedition and while the Chinese desired that their trading partners show due respect to the emperor's authority they were not interested in conquest or occupation. Instead they wished to proclaim the superiority of their emperor and culture to the nations with whom they traded, and the gold and treasures were gifts to impress them.

In 1431 the seventh expedition set sail with a force of more than a hundred large ships. At its return two years later the Chinese had established their hegemony throughout the southern seas from Java to India, Arabia and East Africa. Their achievement was poetically captured in an inscription of sailors' gratitude to a Taoist goddess: "We have set eyes on barbarian regions far away hidden in a blue transparency of light vapours, while our sails, loftily unfurled like clouds, day and night continued their course with starry speed, breasting the savage waves as if we were treading a public thoroughfare." Yet this was the last of the voyages. They had become the subject of infighting at court as traditional Chinese isolationism rebelled against this outgoing policy. Furthermore they were expensive as the Chinese largesse was in no way recompensed by the tribute of subject nations. A succession of imperial edicts threatened harsh punishment for those who ventured abroad. China turned in on itself. The suppression of maritime activity put an end to a tradition of Chinese science and inventiveness which, despite ancestor worship and the reverence for tradition, and unfettered by the demands of a centralized bureaucracy, had marched on uninterrupted for over one and a half thousand years. In Europe, once the home of Greek science and the mighty *Pax Romana*, the picture could hardly have been more different.

An early Chinese mariner's compass. The power of the lodestone, or magnet, has been known in both East and West for over two thousand years. The Chinese first used it for geomancy, the art of aligning buildings and graves in accordance with the "cosmic breath". Later the floating needle was invented, and the technique was adapted to find direction at sea.

A religious image of the Jains of India represents the human body as a model of the universe. The lower part is a series of hells, becoming wider as they reach the feet; the waist in the middle is the Earth; and at the top are the heavens and a small region where liberated souls float in infinite knowledge and bliss.

3

THE GODS RETURN

In the heart of Java a small hill rises out of the luscious, steamy vegetation. It is crowned by one of the most majestic Buddhist relics of Southeast Asia, the *stupa* of Borobudur. This colossal pyramid was built some 1200 years ago and 2.1 million cubic ft (60,000 cubic meters) of stone were needed for its construction. Borobudur is much more than a place of worship and meditation, for this ornate and overwhelming vision of the cosmos symbolizes the Buddhist world and universe. The hill itself evokes Mount Meru, the central apex of the world, which lay beyond the Himalayas and was 84,000 yojanas or 760,000 miles (1.2 million km) high. According to the Buddhist scriptures rivers of sweet water ran in it and on its slopes were beautiful golden houses inhabited by the spiritual beings and their singers and harlots. The side next to our world was encrusted with blue sapphires, their reflection coloring the sky blue. The other sides were studded with rubies and yellow and white gems. Sadly Mount Meru was never found by George Everest.

At Borobudur a terrifying lion's head and aquatic monsters above the main arch signify light and darkness respectively. The foundations were decorated with scenes of the Inferno and then covered as they were not fit to be seen. As you walk up the steps, the four square galleries with their hundreds of reliefs recounting the last life of Buddha on Earth and stories from his previous lives give way to three circular terraces representing the Earth, the atmosphere and the sky. But for Buddhist monks and pilgrims there is a further, more important spiritual dimension. The lower circle is the sphere of desire where individuals must live out life after life as they seek enlightenment. Above is the sphere of forms where, though still tied to the Wheel of Life, they learn to vanquish desire. And then, finally and triumphantly, stands the top circle with its great bell-like *stupa* at the center. This is Araupadhatu, the sphere of non-form, where at last enlightenment overcomes desire and Nirvana is attained.

Buddhism was India's most potent export, and with it travelled a particular world vision portrayed in *wheel maps*. These maps, which perhaps originated in Babylon, were taken by Buddhist scholars in the sixth and seventh centuries AD to China, where Mount Meru became Mount Khun-Lun, the legendary mountain in central Asia or the north of Tibet. Chinese intellectuals, trained in the rectangular precision of Phei Hsiu's grid, looked down their noses at such unscientific incursions.

Nevertheless Buddhist ideas made headway, reaching into Korea and Japan, and it was in the latter country that the most exquisite examples of such maps were made. Unlike the Chinese, the Japanese brought a self-effacing modesty to these beautiful pictures of the world, allowing India and Mount Meru to be firmly in the center and their own islands scarcely visible on the edge.

India was the source of all sorts of wonderful illustrations of Earth and the universe, all emerging from a long tradition of religious storytelling and poetry. Succeeding ideas did not replace each other. Instead they piled up chaotically, leading to an arresting but complicated view of the cosmos in which layer upon layer rose from the infernal to the divine and concentric rings fanned ever outwards. This prompted the English historian Macaulay to deride Indian learning and demand its replacement in native schools with traditional English education. Should the government, he sneered, "countenance, at public expense, geography made up of seas of treacle and seas of butter?" Macaulay was referring to one of the many charming Hindu and Jain world visions, the Sapta-Dvipa Vasumati or Seven-Continent Earth. At the center was Jambudvipa, the rose-apple land where men lived, surrounded by seas of salt, sugar-cane juice, wine, curd and milk.

In some cases the world was a tortoise, its shell the vault of heaven and its underside the Earth. Alternatively the tortoise became a world map, its head, shell and feet demarcating the Earth's regions. A four-continent Earth was sometimes illustrated in the pattern of a lotus leaf with Mount Meru at the center. Then there was the cosmic egg, a representation common to many cultures. The two halves of the eggshell could conveniently be turned into heaven and Earth. Early calculations of the world's size recall Eratosthenes' camels. The distance from Earth to sky was in one version a thousand days' journey by horse, in another the height of a thousand cows standing on one another. Most complicated of all was the Jain universe. The nude human figure, sculpted in gigantic and forthright statues such as Sravanabela-gola, was the most holy symbol of the Jain religion and denoted the rejection of material values. It also represented the universe, with the Earth as the waist in the center, a series of hells below becoming ever wider as they reached the feet, and at the very top a small region where liberated souls floated in infinite knowledge and bliss.

The problem which had so troubled western intellectuals—what held the Earth up?—attracted a variety of ingenious solutions. In one Hindu figure the world rested on the backs of four elephants poised on the shell of a giant turtle which floated on the primeval waters. A nineteenth-century illustration shows the world held up by an Asiatic angel standing on a bowl of rubies supported by a cow standing on a fish swimming in the sea with the sand at the bottom. Entertaining as these fantasies are, the cosmologies of India stemmed from a particular perspective which was also surfacing in the West. The ethical universe displaced the physical and the higher you were the better, as it is with the Western idea of heaven, Earth and hell. The search for a religious and spiritual shape of the world overrode the dogged pursuit of scientific and atheistic exactitude. Religion versus science was the great wrestling match of early Christian thinkers. No one wrestled more vigorously or agonizingly than St Augustine of Hippo. "In him the Western Church produced its first towering intellect—and indeed its last for another six hundred years," wrote his translator, F. J. Sheed. "What he was to mean for the future can only be indicated. All the men who had to bring Europe through the six or seven centuries that followed fed upon him. We see Pope Gregory the Great at the end of the sixth century reading and rereading the *Confessions*. We see the Emperor Charlemagne at the end of the eighth century using *The City of God* as a kind of Bible."

Right An eighteenth-century Indian painting shows "the Churning of the Milk Ocean." The god Vishnu in his tortoise manifestation is seen as the pivot for Mount Mandara, the churning stick. The most common Hindu depiction of the Earth showed Jambudvipa, the rose-apple land where men lived, at the center. The surrounding seas were composed of salt, sugar-cane juice, wine, curd and milk.

What was the world view which colored Augustine's vision? In AD 410 Augustine was fifty-six and enjoying a successful career in the early church. For fifteen years he had been Bishop of Hippo in the Roman province of Algeria. An African citizen of the great empire, Augustine imbibed classical culture at the university in Carthage, immersing himself in Latin literature—Cicero, Varro, Pliny, Virgil—and reading Plato and Aristotle and other Greek writers. From such sources he knew of a round world with the three familiar continents of Asia, Europe and Africa and the strong suggestion of the undiscovered Antipodes to the south. The geography of the later Roman Empire had developed little. A series of itineraries and route maps gave greater precision to military operations and the governance of the empire. In a medieval copy of a Roman route map, called the Peutinger Table after the man who discovered it, the known world is shown in a narrow, elongated strip format. This extraordinary map, made a few years before Augustine's birth, measures 21 × 1 ft (6.4 × 0.3 m). Its ingeniously distorted shape traces staging-posts, spas, rivers and a network of roads totalling 70,000 miles (112,650 km), from Britain to India.

When he was twenty-three, Augustine's ideas underwent a revolution; he was converted to Christianity. The Bible provided new insights, new guidelines, a new

St Augustine, in a fourteenth-century illustration. Augustine, a Christian bishop in Algeria at the time Rome fell to the Goths, thought deeply about the problems which Greek science posed for Christians who wished to believe in the world as depicted in the Bible. Most troubling was Ptolemy's suggestion of a fourth continent to the south, since the Bible held that there were only three continents, one for each of Noah's sons.

cosmos, but also a series of contradictions of the precepts of the now "pagan" literature of the classical world. How, for example, to equate the creation story of Genesis with the physics of Aristotle? In 410 Augustine's world was blown apart. The Germanic tribes, for so long a threat to the northern outposts of the empire, invaded Italy. Rome itself was sacked by the Goths under their king Alaric. Although the empire's hub had been moved east to Constantinople in AD 330, the fall of Rome, the mother of the empire, was the symbol of a culture destroyed, a civilization torn apart. Refugees arrived in North Africa asking why it was that all of Rome's troubles seemed to coincide with the adoption of Christianity as the state religion. The pagan gods had protected the children of the empire for hundreds of years, but now the prohibition of their worship by the Emperors Gratian and Theodosius, promulgated thirty years before the arrival of Alaric, had led to Rome's downfall.

Augustine's *The City of God* set out to refute such speculation and draw its readers to the greater civilization of the God of Christianity. He had to deal with some tricky questions. How, for example, if the peoples of the world were all descended from the three sons of Noah, could the strange races described by Pliny be accounted for? For example, men with one eye in the middle of their forehead or with their soles turned backwards inside their legs; the Pygmies; the females who conceive at the age of five and do not live beyond eight; the Sciapods who lie on their backs in the shade of their one enormous foot; dog-headed men. Some such creatures might be the invention of untrustworthy pagan authors but, believed Augustine, if they existed they were either not human or, if human, they must be descended from Adam and Noah's sons. Like abnormalities at birth, strange races were just part of God's mysterious scheme of things. This was a relief for later mapmakers, who loved to decorate their artifacts with monsters and marvels.

The case of the Antipodes, "men who live on the other side of the Earth, where the sun rises when it sets for us, men who plant their footsteps opposite ours," was more clear-cut. "There is no rational ground for such a belief," proclaimed Augustine. This was a profoundly important declaration which would reverberate down the ages and dictate the shape of medieval maps. After Noah's Ark survived the great flood, the Bible told how God had divided the Earth into three parts, one for each of Noah's sons, Shem, Japheth and Ham. Clearly these parts must be the three known continents and it would be an affront to God to suggest that human beings existed anywhere else in the world. But if they did exist, they could not have been descended from Noah's sons; and that in itself was impossible.

Augustine argued that the proponents of the Antipodes "do not assert that they have discovered it from scientific evidence." Ptolemy himself had acknowledged that the great southern continent was the *Terra Incognita* and that it was probably there just to counterbalance the northern landmass. Augustine covered his options by arguing that even if there was an antipodal continent it did not follow that it was inhabited. As for Noah's descendants having travelled there by sea, Augustine was scathing: "It would be ridiculous to suggest that some men might have sailed from our side of the Earth to the other, arriving there after crossing the vast expanse of ocean." Geographical considerations, however, were tangential to Augustine's overriding purpose of directing his readers towards true understanding and salvation in God. The descriptions of more earthly matters emerged in a series of encyclopedias drawn from classical writing but seeking to place pagan knowledge within a Christian framework. The earliest was written by Paulus Orosius, a Spanish priest and contemporary and acolyte of Augustine. The church was particularly indebted to him as he had brought the relics of St Stephen to the West after their rediscovery in Palestine. Augustine encouraged Orosius to supplement *The City of God* by writing a complete history of the world. Its title, *History against the Pagans*, revealed its true purpose—to demonstrate to the heathen that times had actually been just as bad if not worse before the arrival of Christianity and that it would be a mistake to believe in a long-past golden age benignly overseen by Jove and his fellow gods.

An all-embracing geography was essential so that, in Orosius' words, "when the theaters of war and the ravages of disease shall be described, whoever wishes to do so may the more easily obtain a knowledge not only of the events and their dates but of their geography as well." The intended message was that war and diseases were much worse before Jesus entered the scene. Orosius' description echoes Strabo rather than Ptolemy whose scientific approach took place in relative intellectual isolation at Alexandria, away from the hub of Roman power. The Romans had little interest in theory and Ptolemy made no impact on rulers and the common man. This suited the early Christians well. It was easy to accommodate the consensus picture of the world into the framework dictated by the division between Noah's sons. All sorts of problems would have arisen if Ptolemy's *Terra Incognita* had caught on with others besides the intelligentsia. The clever Augustine could therefore dictate Christianity's response to Ptolemaic theory while more practical writers such as Orosius could repeat the world picture so familiar to the common man at the same time as informing him that it corresponded most satisfactorily with the Bible. Nearly nine hundred years later the creator of the Hereford *mappa mundi*, the climax of medieval cartography, would state that this world map was based on the writings of Orosius and the map of Agrippa.

The works of two other early writers, Macrobius and Martianus Capella, also became a staple part of the medieval curriculum; curious in that neither was a Christian. Their books were intriguing espousals of the classical heritage. Capella's mixture of verse and prose is an allegory about the marriage of Mercury to the virgin, Philologia, whose name means "lover of learning." His last seven books describe the bridesmaids, Grammar, Logic, Rhetoric, Geometry, Arithmetic, Astronomy and Music. The wedding reception was a highbrow affair. Macrobius' contribution was a commentary on Cicero's *The Dream of Scipio*, written by the great Roman orator five hundred years before. Both books were seminal works. "The authors of both were pagans," writes Philip Grierson, "but their Neo-Platonism had many elements in common with Christian philosophy and they in fact dominated Latin cosmological thought throughout the Dark Ages." In Macrobius' *Commentary*, the elder Scipio, whose soul is resting in the heavenly sphere of the departed, describes the universe to his son on Earth: "As I looked out . . . everything appeared splendid and wonderful. . . . From here the Earth appeared so small that I was ashamed of our empire which is, so to speak, but a point on its surface."

Although written over two thousand years ago, this passage, which only survived because Macrobius decided to make it the basis of his commentary, recalls so uncannily the insight offered to the crew of Apollo 8 that one can only admire its leap of imagination. It goes on to relate how below the moon the Earth stands motionless, divided into the classical pattern of climatic zones. All is mortal and transitory except for the souls bestowed upon the human race by the benevolence of the gods. Touchingly, Scipio tells his son how small the part of the world he knows really is and how great are the lands in the other parts of the globe, where his name will never be heard. It was a powerful and prophetic antidote to geographic egocentricity.

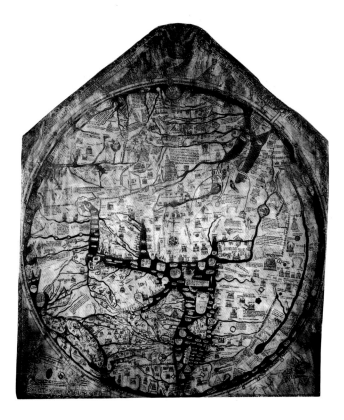

The medieval mappa mundi, *or world map, owned by Hereford Cathedral, in England. The Hereford map was the climax of the Christian search for a world which could accommodate the Bible and Greek and Roman learning. It was a world encyclopedia, including the strange races and animals which Roman writers like Pliny loved to describe, overlooked by the figure of Christ in the east, which was always at the top of such medieval* mappae mundi.

While on the one hand the world of Strabo could conform with that of the Bible, here the Platonic universe, where ideas are real and nature an illusion, could conform with Christian spirituality and its rejection of science. Christian writings are full of admonitions from the early fathers and saints to avoid the pitfall of asking too many questions about the physical world. Augustine, for example, had fulminated against the disease of curiosity. Later St Francis wrote: "Those brothers whom curiosity drives to science will find, on the Day of Judgement, that their hands are empty."

Macrobius' universe could also be viewed as a ladder, the layers of the planets and stars finally leading to the fixed outer layer, the Empyrean, home of the gods and the souls of the departed. "The attentive observer," he wrote, "will discover a connection of parts, from the Supreme God down to the last dregs of things, mutually linked together and without a break. And this is Homer's golden chain, which God, he says, bade hang down from heaven to Earth." This view led to the Christianized hierarchy of the universe which dominated medieval thought. Heaven was above all else, and a descending succession of angels kept the now crystalline spheres on the move, the lower angels being in charge of the moon. Man straddled the borderline between things "corporeal and incorporeal." Trees, plants, stones, insects and worms followed until the bottom was reached. The *Commentary* of Macrobius repays attention because it proves beyond doubt that the world view handed on to Christendom was in all important respects that of Greece and Rome. Augustine, Orosius, Capella and Macrobius himself, whose works would become the basic texts of medieval scholars, bequeathed a sphere. There were disputes over whether the Antipodes existed and, if so, whether they were inhabited. The fallacious notion of an impassable torrid zone was handed on but, as we have seen, this was a long-standing classical tradition. Of Ptolemy's grid and scientific calculations of latitude and longitude no trace remained; his ideas were at that time lost to the West.

How was it then that the first Christian maps were weird rectangular boxes, usually filled with a crude conical mountain, which looked more like a series of jack-in-the-boxes than a plausible version of this world? Their author had once been a widely travelled merchant, a sophisticated man with a broad horizon. He visited western India, where he recorded accurate descriptions of the buffalo and yak. He

St Katherine's Monastery in the Sinai desert, where the extraordinary Cosmas Indicopleustes ("India traveller") retired during the sixth century to write his Christian Topography. *During his travels in India Cosmas could not have failed to notice the wonderful pictures of the universe which stemmed from the Hindu and Buddhist scriptures. When he retired to the desert he decided to shape his world from a selective reading of the Bible.*

watched the caravans of five hundred traders setting off from Abyssinia to buy gold from the natives, who brought it in little lumps from a mysterious source deep in the interior. He sailed to the island of Taprobane (Sri Lanka) where he saw on one of the temples a famous ruby as big as a pine cone. The existence of the stone on the outside of the Buddha Tooth Temple near Kandy was later confirmed by the Chinese pilgrim Hiouen-Thsang. And then this cosmopolitan and urbane man returned to Alexandria, retired to St Katherine's monastery on Sinai and set about constructing a most extraordinary vision of the world.

This man was Cosmas Indicopleustes, the second part of his name meaning "India traveller." Somewhere in his travels in the sixth century his thoughts turned to religion rather than trade and he became a convert to Christianity. His *Christian Topography* displayed all the qualities of fanatical fundamentalism to be found in those who late in life suddenly see the light. His edifice was ingeniously constructed, rising from the foundations of some careful, even crafty, selection from the teachings of the Bible. The venerable historian of the geography of the Middle Ages, Raymond Beazley, wrote of Cosmas: "To demolish false doctrines about the Universe, and to establish the true one, to harmonize science and religion by proving the same things from Scripture and common sense—this was what he set before himself. The aim was thus at starting the same as that of every Christian thinker. It was in the details of execution that Cosmas displayed his surpassing extravagances."

Cosmas' first task was to banish once and for all silly ideas like that of the Antipodes. Here he harked back to an early Christian father, Lactantius, who became tutor to the Emperor Constantine's son. Lactantius, writing at the end of the third century, found the idea of investigating such matters as the size and distance of the heavenly bodies or the shape and nature of the Earth to be completely futile and unproductive. He compared such enterprises with that of trying to assess the character of a city in a very remote country which one has never seen and knows only by name. As for the Antipodes, his verdict was withering: "Can anyone be so foolish as to believe that there are men whose feet are higher than their heads, or places where things may be hanging downwards, trees going backwards, or rain falling upwards? Where is the marvel of the Hanging Gardens of Babylon if we are to allow of a hanging world at the Antipodes?" Cosmas himself rounded on the supercilious air of those who led others into error, insisting that such phenomena as eclipses could only be explained by the sphericity of Earth. Throughout antiquity the sphere, though scientifically proven, must have seemed odd to the average person whose sense of spatial order relied on his senses rather than his intellect. Even a man of letters such as Plutarch had challenged the doctrine of the sphere.

Christian thinkers before Cosmas had compared the universe to a two- or three-story house divided by the firmament and roofed by heaven. To some, such a disputation was best avoided. St Basil declared that religion was not concerned with the shape of the Earth; it did not matter to faith whether it was a sphere, a disc or a cylinder. To Cosmas this view was anathema. These matters could be precisely arranged by a careful reading of the Bible. The clue lay in St Paul's affirmation to the Hebrews that the Tabernacle of Moses, the tent in the wilderness, was the pattern of this whole world "wherein was also the candlestick, by this meaning the luminaries of heaven, and the table, that is, the Earth, and the shew-bread, by this meaning the fruits which it produces annually." The whole was two cubits long and one cubit wide—a rectangle. (A cubit was an approximate measure based on the length of the forearm.) Manuscript illustrations of Cosmas' world show it as a vast box, like a travelling trunk. Its bulging lid is the vault of heaven from where the Creator is seen

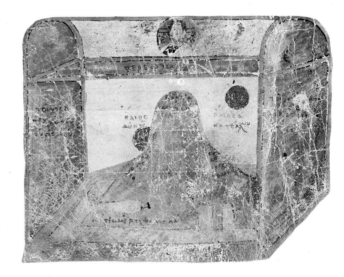

The world according to Cosmas. He took the idea from St Paul's statement that the Tabernacle of Moses, the tent in which the prophet lived in the wilderness, was a model of the whole world. Cosmas thus constructed a rectangular Earth like a vast box, its bulging lid the vault of heaven from which God surveyed his creation below. To the north was an enormous mountain around which the sun moved, its shadows causing night to follow day.

surveying his works below. Dominating everything is the great mountain to the north around which the sun moves, the obstruction which causes night to follow day.

Cosmas' phantasmagoria covers a wide range of natural phenomena. It explains, for example, why the sun must be only two fifths the size of the Earth; why only two heavens can exist as opposed to the nine spheres of the heathen; why there can be only one face of the Earth; why the land is higher and the sea deeper in the north and west than the south and east. His theories fill twelve long volumes, every assertion accompanied by the most detailed proof. Cosmas' work was a unique and appealing, if somewhat bizarre, manifestation of a total vision. Whereas Ptolemy had constructed a world from his hundreds of coordinates, whose scientific presentation obscured their shaky foundations, Cosmas employed the coordinates of biblical teaching. Was the conversion from merchant to visionary simply inspired by God and the Bible? Or, perhaps, did Cosmas during his travels encounter the exotic visions of India, the universes of Hindus, Jains and Buddhists? And did he decide that such a construction as they had made from their religious traditions was owed to Christianity and the Bible? There is an overpowering parallel between Cosmas' great mountain to the north and Mount Meru. The motivating force, the wish to build such complexity from a literal accordance with religious teaching has more in common with the traditions of the East than the West.

The power and uniqueness of Cosmas, and particularly his authorship of the earliest surviving Christian maps, persuaded many to believe that, after him, a flat Earth returned to dominate Christian thought. In fact his influence in Europe was negligible. It was not mentioned by medieval commentators, with one exception, who wrote dismissively that "the style is poor, and the arrangement hardly up to the ordinary standard.... he may fairly be regarded as a fabulist rather than a trustworthy authority." Cosmas was entertaining, ingenious, almost breathtaking in his translation of the Bible into physical terms. In the sixth and seventh centuries, the darkest period of the Dark Ages, the mainstream tradition of Western Europe found its sole standard-bearer in another great encyclopedist, Isidore of Seville. Isidore also left behind a world map, this time clearly if simplistically rooted in the three known continents of classical orthodoxy. It is based on the T–O figure which would dictate the shape of the great majority of medieval *mappae mundi*. The O is the circle of the world and the T is the three waters, the Mediterranean, the Nile and the Don, which divide it into the three continents of Europe, Asia and Africa.

Isidore was a more intriguing and influential character than the crude outline of his world view might suggest. He straddled the sixth and seventh centuries and had what was at that time the advantage of living in Spain. Italy had taken the brunt of the barbarian invasions from the north, its cities and civilization fractured in the first swathes of destruction. Spain, although it too suffered greatly, tended as the western outpost of the Roman Empire, to see these early invaders only after their numbers had been depleted and their energy and savagery reduced by the exertions of their initial conquests.

Isidore succeeded his brother Leander as Bishop of Seville in about 600 and died there some thirty-six years later. Above all his function was to give a reassuring and complete picture of the world to the faithful in difficult times. The works of Pliny, Orosius, Macrobius, Augustine, Lactantius and many other Roman and patristic writings (Ptolemy was temporarily forgotten) coalesced in Isidore. He was essential reading for students, priests and mapmakers for hundreds of years. Given his influence, Isidore's idea of the precise shape of the world has excited heated debate. He was in no doubt about the universe itself: "The sphere of the heavens is rounded,

This simple map was the compromise between the Bible and classical learning reached by Isidore, the seventh-century archbishop of Seville. East was at the top and the three continents of Europe, Africa and Asia were divided by the Mediterranean Sea and the rivers Nile and Don. But Isidore was more interested in intriguing mysteries like the exact location of Paradise, and whether the stars possessed souls.

and its center is the Earth, equally shut in from every side." But in the case of the Earth, Isidore seemed more confused. For example, his description of the climatic zones located them very curiously. Instead of their being parallel bands running round a spherical Earth, he illustrated them as flat circles filling a flat disc-like Earth, like circular light reflectors mounted on a bicycle wheel. Even more oddly, he made the north and south circles adjacent, since being "rendered waste by the great rigor of the climate and the icy blasts of the winds," they would need to be at the furthest possible distance from the sun. As for the Earth itself, "the circle of lands [*orbis*] is so called from its roundness, which is like that of a wheel, whence a small wheel is called *orbiculus*. For the Ocean flowing about on all sides encircles its boundaries." No one can deny that a wheel is flat but is it possible that Isidore was simply describing the known world rather than the whole Earth? Such an interpretation becomes possible when we read Isidore's rather daring statement that there might be a fourth part of the world across the Ocean "which is unknown to us on account of the heat of the sun, in whose boundaries, according to story, the Antipodes are said to dwell."

The verdict on Isidore must be left open but his was at least a realistic view compared with the eccentricity of Cosmas. Sphericity and the Antipodes continued to worry Christian thinkers. Despite Cosmas there were references in the Bible to the "circle of the earth." On the other hand the gospels stated that the Lord "shall send his angels with a great sound of a trumpet and they shall gather together [his elect] from the four corners of the earth." A circle has no corners; a square does. One ingenious solution was to place the square inside the circle or even the circle inside the square. The Venerable Bede, the greatest figure in English letters at the turn of the eighth century, blended common sense and the Bible: "The cause of the unequal length of the days is the globular shape of the Earth, for it was not without reason that the Sacred Scriptures and secular letters speak of the shape of the Earth as an orb." By early medieval times the sphere was taken for granted by Christian writers and the orb became a frequent symbol in art.

Isidore's lack of clarity on such a fundamental matter seems surprising. Yet, like many Christians, he had little interest in immersing himself in a question merely relating to the physical world, for his preoccupations were metaphysical and spiritual. He delighted his readers with appealing comparisons between the human body and the four elements: fire, air, water, and earth. The head was the heavens, its eyes the sun and moon; the breast with its regular breathing, was the air; the belly, as the collector of all the humors, was the seas; the feet were dry like the earth. Then there was the allegory of the universe: the sun was Christ, the stars were the saints, thunder a rebuke from on high. This raised the intriguing question of whether the stars had souls.

The greatest game of all was the search for Paradise. Isidore was not the first to speculate on its location but few matched him for precision. "Paradise," he opined, "is a place lying in the parts of the Orient whose name is translated out of the Greek into the Latin as Hortus. In the Hebrew it is called Eden, which in our tongue means delight." Isidore described this wonderful place in detail. There was a continual spring temperature, every kind of tree, particularly the Tree of Life, and a great spring which divided into the four rivers which flowed from Paradise. There was only one problem, caused by Adam and Eve: "Approach to this place was closed after man's sin. For it is hedged in on every side by sword-like flame."

The earliest and most famous navigator to set sail for Paradise was the Irish monk St Brendan. His voyage has excited much speculation that Irish sailors did

The Venerable Bede, whose writings at the turn of the eighth century applied common sense to the Bible. The biblical world picture was open to many interpretations: there were references to the "four corners of the Earth," but also to "the circle of the Earth." Bede was in no doubt and wrote of the "globular shape of the Earth." By early medieval times the sphere was taken for granted by Christian writers and the orb became a frequent symbol in art.

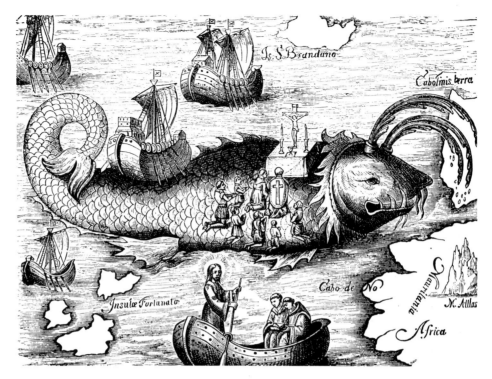

The mythical voyage of the Irish monk, St Brendan, discoverer of the earthly paradise known as the Garden of Eden. According to the stories, St Brendan headed into the Atlantic and encountered all sorts of fabulous islands and sea monsters, including a whale on which he held communion. Eventually Brendan found Paradise, and his island became an established feature of medieval and Renaissance maps.

indeed make a series of expeditions from the rugged west coast of their native land, encountering whales, icebergs and volcanoes and visiting such places as the Faroe Islands and Iceland. Brendan's own voyage, chronicled three centuries later, took in a succession of amazing sights. On one island his band of monks found a tree with a colossal trunk full of pure white birds who confessed to Brendan that they were fallen angels and sang hymns in unison to seek redemption. On a solitary rock in the middle of the sea, battered by waves like a skiff in a whirlpool, the swarthy and disfigured shape of a desperate man was revealed to them. It was Judas Iscariot. Finally Brendan and his brothers found themselves enveloped in a thick darkness which suddenly cleared to reveal a great land brilliantly illuminated. A young man approached and said that this was the place they had been seeking. A vast river flowed through it and even after forty days of wandering they did not reach the furthest shore. Having filled their ships with precious stones they sailed home and Brendan died shortly afterwards. Brendan's island became a familiar feature on charts of the Atlantic for the next thousand years. No wonder legends arose of an Irish discovery of America.

Paradise became an essential part of medieval maps, which were illustrated with Adam and Eve standing either side of the Tree of Knowledge, sometimes with the four rivers emerging from it. (Cosmas' illustration of Paradise was one of his more believable efforts.) The alleged discovery and location of Paradise was an important selling point of medieval travelogues, which were full of claims and counter-claims. Christopher Columbus was even to suggest a Paradise so remarkable that it required his fellow beings to change the shape of the world: "I have now seen so much irregularity that I have come to another conclusion respecting the earth, namely that it is not round as they describe, but of the form of a pear . . . or like a round ball, upon one part of which is a prominence like a woman's nipple, this protrusion being the highest and nearest the sky, situated under the equinoctial line, and at the eastern extremity of this sea where the land and the islands end."

Some interpretation is required. Columbus was at the Orinoco River. Its rush of fresh water was so great that it seemed it could only be flowing from Paradise. He believed that he was at the easternmost point of the world, somewhere beyond China. He saw a great landmass unexplained by Christian, let alone Ptolemaic, geography, stretching into the distance. Here then was an extra piece of land which jutted off the world. Paradise, so far undiscovered but known from the writings of Genesis to exist, was the logical explanation. Curiously, modern technology has shown that the Earth is very slightly pear-shaped, the distance from the equator to the South Pole being marginally greater than to the North Pole. However, any resemblance between this bulge and Columbus' protuberance is purely coincidental.

But the best Paradise story dates from the nineteenth century. It is a testament to human unwillingness to accept symbolic rather than literal truth. General Charles "Chinese" Gordon was a brilliant, flawed Victorian eccentric, hero and victor in the Taiping rebellion, the martyr to Khartoum, a heavy drinker, a latent homosexual, generous, explosive, conscience-stricken and, in religion, deeply fundamentalist. In 1881 he accepted a post in Mauritius where he became interested in botany and found unaccustomed time for his Bible studies. He was ordered to check harbor installations at Mahé, the largest island of the Seychelles. While there he visited the neighboring island of Praslin which, with a dazzling display of evidence, he declared beyond doubt to be the long-lost terrestrial paradise. The key to this startling assertion was the *coco de mer* palm tree. It has a male and female variant, umistakable from their sexually suggestive fruits. The male fruit is phallus-shaped and the female is, as Gordon put it, "shaped in the husk like a heart, when opened like a belly with thighs." Here was the Tree of the Knowledge of Good and Evil whose fruit (a double coconut in fact) Eve had been tempted by the serpent to eat. The fruit's symbolism added power to the sexual desires released by Eve's fall.

Further evidence followed. Gordon claimed to have seen a serpent, while the breadfruit tree, which also flourished on Praslin, he identified as the Tree of Life which stood next to the Tree of Knowledge in the Garden of Eden. Finally, and most ingeniously of all, he decided that the Bible needed to be reinterpreted to allow the

In 1881 the English soldier, Charles Gordon (above), famous later as hero of Khartoum, accepted a post in the Seychelle Islands in the Indian Ocean. There he found time to combine two areas of expertise, mapmaking and interpretation of the Bible, in his search for Paradise. The fruit of the coco de mer palm tree (left) provided Gordon with a vital piece of evidence that the island of Praslin in the Seychelles was the true location of Paradise. The male fruit is phallus shaped and the female is, as Gordon put it, "shaped in the husk like a heart, when opened like a belly with thighs." These fruits were clearly the sign of the fall of Adam and Eve.

four rivers of Paradise to flow into one great basin rather than from it. The basin he found west of the Seychelles deep in the Indian Ocean where the water reaches the great depth of 2600 fathoms (4750 m). The four rivers were trickier. The Tigris and Euphrates, both mentioned in the Bible, accounted for two. By reconstructing the world as it was before the Flood, Gordon surmised that the River Nile had once discharged straight into the Red Sea (from which it could meander on to its basin near the Seychelles) to which it was now connected by an ancient Greek canal. The fourth was found in a now dried-up river bed south of Jerusalem which reached the ocean via an underground canal. Like Cosmas' scheme, it was wonderful, crazy, clever, and it was illustrated by a series of maps expertly drafted by Gordon.

Gordon's finely drawn sketches are far removed from the square, boxlike Paradise which appears in the maps of the Spanish Benedictine abbot Beatus. In the late eighth century Beatus wrote a popular commentary on the Apocalypse of St John and accompanied it with a map. There is one interesting innovation: Beatus wished to illustrate the mission of the apostles to all four corners of the Earth, and either from a desire for symmetry or through intellectual conviction he concluded that there was a fourth continent and that it was inhabited. A legend on the map says: "We are told that the Antipodeans, around whom revolve many fables, live within its confines." Christian doctrine on this matter was in a state of flux. A contemporary of Beatus, Virgil, Bishop of Salzburg, was hauled before the Pope for suggesting that there existed another inhabited world. The Pope gave Virgil the benefit of the doubt on the grounds that he may have misunderstood earlier writings.

The combination of classical knowledge and Christian preoccupations found its most beautiful and detailed encapsulation in a succession of medieval *mappae mundi*. Sometimes these could be tiny; the famous Psalter Map in the British Library, a richly colored illustration in a thirteenth-century manuscript, measures $5\frac{3}{5} \times 3\frac{3}{4}$ in (14.3 × 9.5 cm). It is probably a copy of the world map made for Henry III's audience chamber at Whitehall, evidence itself of a burgeoning curiosity about the world at the highest level stimulated by the renaissance in Europe of the eleventh and twelfth centuries. The Psalter Map's miniaturized features are unmistakable. East is at the top with the Garden of Eden. Above, Christ, with angels on either side, surveys the world. To the far northeast can be seen the threatening tribes of Gog and Magog, who were held back only by a wall of iron built by Alexander the Great. These dreaded servants of the Antichrist were expected to erupt on the Day of Judgement and wipe out Christendom. Roger Bacon, the rationalist light of medieval science, pressed for detailed scrutiny of their exact location so that Christians could plan against the coming invasion.

Sometimes these maps were enormous. The Ebstorf map, named after the Benedictine monastery in Germany where it was discovered, measures almost 11 ft (3.4 m) square. Sadly the original was destroyed in the bombing of Hanover in the Second World War, but the faithful reproduction now in the convent at Ebstorf shows how the world has become the body of Christ. His head is at the top, the rivers are his veins, his feet are at the south and his hands point out from east and west. Its author intended his great work to be more than a religious symbol. An inscription in the top-right corner reads: "It can be seen that [this work] is of no small utility to its readers, giving directions for travellers, and the things on the way that most pleasantly delight the eye."

Most famous of all such maps is the Hereford *mappa mundi*. In 1989 it suddenly became the subject of great controversy when the cathedral authorities, who own it,

The medieval world map in a thirteenth-century book of psalms, a treasure of the British Library in London. This was a miniature copy of a great map which hung in the chamber of King Henry III of England. Of all the surviving mappae mundi, *it is perhaps the most beautiful, being richly colored and exquisitely detailed.*

A page from the medieval road atlas created by Matthew Paris, a monk who lived in St Albans, north of London, in the thirteenth century. This strip map for pilgrims started at the old St Paul's Cathedral in London and continued across the English Channel, and down through France to south-eastern Italy. From there the final stage of the journey to Jerusalem would be made by boat.

decided to sell it to raise funds for restoration. Protests flooded in as people began to realize that the map was a vital part of both British and European culture. The sale was cancelled. The map was designed at Lincoln in the 1280s by Richard de Bello, a prebendary of Haldingham near Lincoln, and probably painted by professional illuminators. When Richard later moved to Hereford, he took it with him. The Hereford map brings together most of the geographic and cartographic traditions of the previous thirteen hundred years. In the words of Peter Barber of the British Library: "The map was primarily intended to show all of God's creation in its proper historical and geographic setting: a visual affirmation of traditional learning and faith. It is also an *estoire* locating events in human and divine history, from the fall (Adam and Eve at the top or east) to the thirteenth century—and beyond to the Last Judgement, shown above the map." Geographic accuracy took second place to a visual encyclopedia of the world. Contained within it are the strange races of Pliny, whose existence Augustine had been at such pains to explain; the legends and marvels of India and Africa; and intriguing animals such as the unicorn. The Hereford map also gave up-to-date information on important towns and trade and pilgrimage routes.

Richard de Bello, as he composed his great artifact, would have known the work of Matthew Paris, the monk of St Albans and chronicler of the middle years of the thirteenth century. He only once left his country, when he was sent on a mission by Pope Innocent IV to reform a dissolute monastery in Norway. Owing to the idleness of its monks, its religion had disappeared, its fame had dwindled and its goods had been dissipated. Paris put all this right and returned to his books at St Albans where he was writing the *Chronica Majora*, a detailed history of his times. He never left England again. Despite his sedentary life, a wide range of contacts enabled Paris to keep up with European news. He was in frequent contact with King Henry III which makes it all the more remarkable that his writings should contain such vitriolic attacks on the king as "a vigilant and indefatigable searcher after money." Paris himself had an almost unnatural interest in money and viewed all taxation as extortion. His contacts and reading aroused a fascination with geography, particularly that of the Holy Land. He drew relatively accurate maps of Britain and Palestine; Jerusalem had become the center of world maps and pilgrims were always eager for information on the Holy Land. Paris' most original contribution was a route map of the journey from London to Apulia in southern Italy. It is a highly illustrated and descriptive antecedent of today's route maps.

In this profusion of medieval mapmaking one important ingredient was missing: the scientific framework of Claudius Ptolemy, whose influence had simply disappeared from the Christian world. His saviours were the Arabs. The torch of Islam, ignited after the death of Mohammed in 632, had over the next hundred years swept through Mesopotamia, Palestine, Egypt, North Africa and part of Spain. The all-conquering Moslems brought with them from the Arabian desert a hunger for new learning and culture. They became the heirs to the philosophy of Aristotle, the medicine of Galen, and the astronomy of Ptolemy. In the ninth century the Caliph founded in Baghdad an academy called the Hagia Sophia, the Greek for Sacred Wisdom. The major Greek works were translated into Arabic.

Arab astronomers sought to improve and revise Ptolemy's calculations, with considerable success. The latitude of Baghdad for example was measured by astrolabe as 33° 20′ – correct to within one minute. Attempts were made to calculate longitude from eclipses of the moon. Ptolemy's *Geography* was quoted in itineraries usually written as a by-product of the knowledge gained from pilgrims who descended on Mecca from all parts of the world. Most of this Arabic scholarship was purely derivative. Occasionally, inquiring minds questioned traditional orthodoxies. The tenth-century writer, Al-Masudi, for example, rejected Ptolemy's idea of a land link between the tip of Africa and southern Asia which enclosed the Indian Ocean. Arab maps, though the product of science rather than religion as in Christendom, showed little advance on classical achievement. Their most notable exponent was Al-Idrisi who settled at the court of the Norman King Roger of Sicily in Palermo. The city was a world crossroads where merchants and sailors from the twin worlds of Islam and Christendom exchanged a breadth of world knowledge. Al-Idrisi agreed to write for King Roger, an inquisitive and appreciative patron, a geographical treatise winningly described as an "amusement for him who desires to travel round the world." It took the form of a travelogue illustrated by a world and then a series of regional maps. The technique was based on Ptolemy, even if some of the conclusions were different. For example, Al-Idrisi stated that the other end of the Nile flowed towards the west coast of Africa whereas Ptolemy diverted it to his invented Mountains of the Moon. Curiously, Al-Idrisi's work had little impact in the rest of Europe. The lives of Christendom's medieval scholars had been transformed by the rediscovery of classical thought, which had been transferred west from the Arab world. Aristotle was the new king of Academe. But while perceptions of the universe were initially revolutionized, they later became stultified by adherence to Aristotelian physics.

Geographic knowledge was improved by the crusades and pilgrimages but representations of the world remained locked in the inflexible pattern of the *mappae mundi*. The shackles were to be loosened by two acts of liberation. At the turn of the thirteenth century a monk called Maximus Planudes was accustomed to scour the secondhand bookshops of the bazaars of Constantinople for the works of the great classical writers. He once restored an old copy of Strabo's *Geography* which had been gnawed by rats. Planudes came to realize that the key classical text on geography had been written by Ptolemy, and set his heart on finding one. Eventually he found a copy of the text with all its coordinates but no maps. He commissioned local artists to produce new maps according to Ptolemy's instructions. Planudes' find became the talk of Constantinople. A beautifully ornate copy was made for the emperor. However, it was another hundred years before this rediscovered knowledge made its way to Western Europe. In 1395, as Constantinople squared up to the threat of the Ottoman Turks, a diplomat and scholar called Manuel Chrysoloras was sent to northern Italy to enlist the help of Venice and Florence. He took the emperor's presentation copy of Ptolemy's *Geography* as a gift. In 1406 it was translated by Jacopo d'Angelo, who dedicated his work to Pope Gregory.

As European horizons expanded in the fifteenth century Ptolemy's *Geography* became the yardstick against which the results of exploration and new discoveries were judged. Updaters of his work were embarrassed by the need to interfere with the great Alexandrian's coordinates. But ironically, as Ptolemy reached the height of popularity in the fifteenth century, achieving a fame towering over that accorded him by his dull Roman contemporaries, it was becoming clear that his work, however brilliant the calculations and method, rested on shaky factual foundations. Originating in the Mediterranean, a totally new kind of map, the sea chart, was beginning to record more accurately than ever before the outline of the Earth.

Chora monastery in Istanbul, home of the monk Maximus Planudes, who rediscovered Ptolemy's Geography *at the end of the thirteenth century. Planudes scoured the bookshops of Constantinople for the works of the ancients, and the* Geography *was his greatest find. Later it was to become a key work for Renaissance scholars.*

4

SECRETS
OF THE
EAST

I n 1291 two Genoese brothers, Ugolino and Guido Vivaldo, embarked on the most remarkable maritime undertaking not only of their own time but also of the next hundred and fifty years. With their two galleys, the *Allegranza* and the *St Antonio*, a pair of Franciscan missionaries, and a complement of north Italian warriors and traders, they aimed to find a route by the Ocean Sea round Africa to India and the East. Their prize would be a direct link to the silk, spices, silver and many other Eastern luxuries which so aroused the appetite and greed of Europe. The exact design of the Vivaldos' ships is unknown but if they were standard Genoese galleys heavily manned with over one hundred oarsmen the breathtaking ambition of such a voyage through the sapping heat of the equator down to the tip of Africa many thousands of miles to the south, may today seem foolhardy, even suicidal. Yet, if we sweep aside modern familiarity with the world and imagine the prospect beckoning a courageous thirteenth-century navigator from the Mediterranean, the Vivaldos' voyage was not merely a rational enterprise; it was prophetic.

By the time of the Vivaldos the East was less remote, the idea of a linked world more plausible. After the confines of the Dark Ages, the Crusades had been the engine of a reawakening of the European world view. Knowledge of the East, in particular China, had been minimal between the seventh and twelfth centuries. Then rumors arose to whet western curiosity. The most potent of these was the alleged presence of a Christian king called Presbyter, or Prester, John who ruled over a great empire and would come to the rescue of the Christian states in the Near East which had been established by the First Crusade but were now beleaguered by Saracen counterattacks. The author of this legend was Bishop Hugh of Jabala, today the Lebanese port of Jablah. In 1145 he was despatched to the West by the Crusaders to seek Christian reinforcements. In November of that year he met the Pope at the central Italian town of Viterbo. Their conversation was preserved for posterity by one of the most reliable of medieval historians, Otto, Bishop of Freising in Germany.

Bishop Hugh spoke of a certain John, a king and priest "who lives in the extreme Orient, beyond Persia and Armenia, and is a Christian although a Nestorian." (Nestor, patriarch of Constantinople from 428 to 431, taught that Christ's divinity and humanity were not united in a single self-conscious personality.) John had

Left *Arabia and the eastern half of Africa, from a 1558 atlas by the Portuguese Diogo Homem. The legendary emperor of the East, Prester John, who had gripped the imagination of Europeans for 400 years, has here been relocated to Africa, and is shown enthroned in the region of modern Ethiopia.*

Genghis Khan the Mongol warrior (top) *swept south into China in 1206 and founded a new dynasty. His conquests created a vast empire stretching from the Yellow River to the banks of the Danube and from the Persian Gulf to Siberia. From 1250 to 1350 it was possible for Europeans to travel overland to China. Kublai Khan* (above) *was the emperor whom Marco Polo served.*

recently defeated the Persians in bloodthirsty battle and had headed west to come to the aid of the Church in Jerusalem. "When he had come to the river Tigris," Bishop Hugh continued, "he had not been able to take his troops across it in any vessel. Then he had turned to the north, where, he had heard, the river sometimes froze over in the winter cold. He had tarried there for some years, waiting for the frost, but on account of the continued mild weather there was very little, and finally, after losing much of his army because of the unaccustomed climate, he had been forced to return home." Bishop Hugh concluded by adding that Prester John was descended from the Magi, the Three Wise Men, and that he ruled over their lands. The story was tantalizing if improbable. Bishop Hugh's story corresponded to an historical event, the defeat in 1141 of the Persian Seljuks by the Khitans, a tribe who had once ruled the north of China but were then driven west by a new dynasty and clashed with the Seljuks near the Silk Road town of Samarkand. Prester John's identity therefore stemmed both from the Khitan prince, Yeh-lu Ta-shioh, and from the tradition of Nestorian Christianity in India.

Twenty years later the legend was strikingly confirmed in a letter addressed by Prester John to the Byzantine emperor Manuel I. Its glowing account of the amazing lands under his sway became one of the most popular documents of the time. Nearly one hundred manuscripts of the Latin version alone survive to this day, a high survival rate compared with that of other medieval writings. Prester John portrayed a truly glittering prospect. "Our magnificence dominates the Three Indias," he wrote, "and extends to Farther India, where the body of St Thomas the Apostle rests." Prester John claimed the allegiance of seventy-two provinces, whose kings paid tribute to him. The abundance of his domains was staggering: "In our territories are found elephants, dromedaries, and camels, and almost every kind of beast that is under heaven. Honey flows in our land, and milk everywhere abounds. In one of our territories no poison can do harm and no noisy frog croaks, no scorpions are there, and no serpents creep through the grass. No venomous reptiles can exist there or use their deadly power. In one of the heathen provinces flows a river called the Physon, which, emerging from Paradise, winds and wanders through the entire province; and in it are found emeralds, sapphires, carbuncles, topazes, chrysolites, onyxes, beryls, sardonyxes, and many other precious stones."

Unfortunately Prester John's letter was a fabrication. Its likely author was a western European monk well versed in the geographic literature of his time. It has proved impossible to determine his identity or purpose. Perhaps he intended to stimulate Christian interest in the East, or to provide a literary entertainment, or to produce a description of a Utopian land to set against the evident degeneracy of Europe. Perhaps it was just a prank. Its first translator, Roau d'Arundel, wrote in a prologue that its chief value lay in its revelation of "the great miracles of the Orient." The details did not have to be swallowed verbatim. Nevertheless Prester John and his kingdom captured the imagination of medieval travellers, who soon went in search of him. The impetus was the Mongol conquest of China by Genghis Khan. He and his successor, Kublai Khan, created a vast empire stretching from the Yellow River to the banks of the Danube, from the Persian Gulf to Siberia, through which it was possible for Europeans to travel uninterrupted. Predictably, Genghis Khan was identified with Prester John but this idea was quashed as soon as the first Christian missionaries made contact with the new emperor.

For a single century, from around 1250 to 1350, the gateway to the East was open. The Mongols, or Tartars, had terrified Christendom with their rapid advance through Poland and Hungary; a new crusade was even proclaimed against them. But

Marco Polo, the Venetian traveller who served at the court of Kublai Khan in China. He excited the West with his stories of the fabulous East, describing the emperor hunting with 20,000 huntsmen and 10,000 falconers. There was also a commercial lure: in the port of Zayton the explorer saw a hundred times more pepper on the quayside than in the largest western entrepôt of Alexandria.

this was unnecessary when, with dominion over all Europe beckoning, the Mongol armies were summoned back to Asia after the death of the great Khan Okkodai. Instead the crusaders found in the Mongols unsuspected and unintended allies as they swept through the Saracen lands of Syria and Persia, removing the Caliph of Baghdad along the way. Franciscan friars headed east through the newly pacified and regulated lands in the hope of converting the Khan to Christianity. They found an empire tolerant of many religions, with Buddhists, Muslims, and Manicheans all holding out the prospect of salvation to the Mongol leader. But with his civilized empire on Earth, he had little need of such solace. While the monks had limited success and Prester John was nowhere in sight, the commercial attractions of more direct links with the East were increasing with every report. When the Vivaldos set out, Marco Polo was serving in the government of Kublai Khan. His description of China and its magnificent cities such as Kin-Sai, preeminent in the world in grandeur and beauty, ignited the imagination of the Europeans.

Marco Polo described not just the emperor who went hunting with 20,000 huntsmen and 10,000 falconers but the ports such as Zayton, where "the quantity of pepper imported . . . is so considerable that what is carried to Alexandria to supply the western parts of the world is trifling in comparison, perhaps not more than the hundredth part." The problem for Genoa was the constant battle she had to fight with her arch-rival Venice for control of the ports of Egypt and the Levant, through which the most exotic and profitable products of the East were traded. The long-term solution, which foreshadowed the achievements of the Portuguese two hundred years later, was a direct sea passage to the East. Genoa, more than any other city, broke the deadlock of the oceans. Its harbor is ringed by a steep curtain of mountains and even in medieval times its populations crowded into a squash of tall buildings along its narrow strip of coast. It was a city-state of bankers, merchants and pirates governed by a commune of families and factions which spent most of their time fighting each other. In one forty-year period there were no less than fourteen revolutions. Families such as the Grimaldis, who still rule Monaco, had to find new land along the coast or on islands including Sardinia and Corsica. Unlike Venice, Genoa had no hinterland to command and tax; the only domain into which it could expand was the sea. Genoese sailors were said to be unequalled in their skills and zeal for profit and it is hardly surprising that Columbus was a native of that city.

We know how the world looked to a medieval monk constructing his *mappa mundi*. But what was the prospect for a sailor whose very life depended on an accurate assessment of the unknown? Medieval Europe still saw the Earth as it was outlined in the *mappa mundi*. The three-continent structure with a surrounding ocean was firmly entrenched. (Ptolemy's idea of a landlocked Africa had not yet returned to confuse matters.) Anyone who read the Arab geographers would have had nightmares about the Atlantic Ocean. Al-Idrisi, the most sophisticated Arab mapmaker, wrote at length about the thick and perpetual darkness which brooded over what was then referred to as the Western Ocean. Interpreters of the Koran declared that a man insane enough to embark upon its waters should be deprived of his civil rights; but as, according to such strictures, he was bound in any case to lose his life, the penalty was perhaps superfluous. As late as the fifteenth century another Moslem detractor wrote that the Western Ocean was "boundless so that ships dare not venture out of sight of land, for even if the sailors knew the direction of the winds, they would not know whither those winds would carry them. As there is no inhabited land beyond, they would run the risk of being lost in mist, fog or vapor."

These fears were no more than speculation, yet Arab sailors did not dare to venture far into the Atlantic. To the Genoese and other Italian traders the picture gained from experience was less daunting. The passage from the Strait of Gibraltar round to northern Europe was an important trade route, and, although the filthy weather of the North Atlantic must have sent many shivers down the spines of drenched and frozen sailors, the fog always cleared and Al-Idrisi's "perpetual" darkness always seemed to lift. By this time Genoese ships regularly put in to ports on the northwest coast of Africa beyond Tangiers with no apparent difficulty. The first tentative steps seemed to prove that the Atlantic was as navigable as any other sea. Whether or not the Genoese knew about the Greek and Phoenician expeditions beyond the Pillars of Hercules (the Strait of Gibraltar) is not known but at the very least their sophisticated maritime culture was unlikely to be disturbed by superstition, particularly when it came from the Moslems whose ships were the favorite targets of Genoese piracy.

The unknown reaches of the Atlantic were the first traditional barrier facing the Vivaldos; the second was antiquity's firmly entrenched belief in an impassable torrid zone. The latter was a vital question for anyone sailing south, although here again there were grounds for optimism. First, the size of Africa was still a matter of guesswork; it might be relatively small and the journey around it need not necessarily reach as far as that unspeakable equatorial zone. Secondly, a few theorists of the thirteenth century had begun to question the rationale behind this notion of the torrid zone. The Dominican friar Albert the Great realized from the evidence of medieval travelogues and even classical accounts that the presence of cities in southern India and of Ethiopian tribes near the equator proved that the region "which seemed to the Ancients to be a Torrid Zone is habitable."

Albert provided further encouragement with his analysis of the vexing question of the Antipodes, the men who lived on the other side of the world. Why was there no evidence of contact between this side and theirs? Rumor had it that mountains of magnetic rock stood between, drawing back those who tried to pass. Albert was too clever to be taken in by such a hoary old tale: "It seems to me incredible that they should be impassable everywhere. I believe it is truer to say that the crossing is difficult but not impossible, and this, on account of vast deserts of sand rendered waste by the fierceness of the sun; this makes the undertaking impossible unless provision is made for a long journey. It is for this reason, I think, that there is little or no intercourse between the men who live beyond the Equator in southern climes and those who live with us in the northern hemisphere." This was excellent news for the Genoese brothers, for it meant that there was no danger of burning and shrivelling up in a molten heap as you crossed the equator, although hats and plenty of supplies would be necessary.

The technology for such an enterprise was becoming available. At the end of the twelfth century the compass had arrived in the Mediterranean lands, and all sorts of delightful stories and superstitions had grown up around it. The London doctor, William Gilbert, in a treatise called *De Magnete* written three hundred years later, recalled, if only to dismiss, these charming properties of the lodestone: "If pickled in the salt of a sucking fish there is a power to pick up gold which has fallen into the deepest wells." The compass was said to have "the power to reconcile husbands to their wives, and recall brides to their husbands," a useful instrument for sailors whose wives sought comfort while they were away at sea. Gilbert went to drastic lengths to disprove another popular notion, namely that onions and garlic obliterated the power of the lodestone and compass needle. "When I tried all these things I found them to be false: for not only breathing and belching upon the Loadstone after eating of Garlick, did not stop its vertues: but when it was all anoynted over with the juice of Garlick, it did perform its office as well as if it had never been touched with it." History does not record whether Italian and French sailors, ignorant of Dr Gilbert's experiments, were forced to forgo this most important part of their national diet for the sake of safety at sea.

Contemporaneously with the compass, the first portolan charts began to appear in the Mediterranean. These sea charts would be a revolutionary new force in shaping man's vision of the world. Not only did they become an important record of discovery, but they were also a vital storehouse of knowledge; so critical that the Spanish and Portuguese would later make them state secrets the transmission of which to rival nations could be punished by death. The appearance of portolan charts could hardly be more different from that of the medieval world maps or even the route maps of such as Matthew Paris. They were designed for practical, indeed life-saving, use. They were a graphic rendering of written descriptions which had their origin in the classical periplus (a description of seas and coastlines). Beautifully outlined in hand, the coastlines were precisely mapped from the evidence of firsthand experience and navigational drawings. The main ports and coastal features were illustrated. The degree of embellishment varied; thousands of them that were purely practical and no longer survive, must have been without adornment. In some that do survive, the illustrations of cities and, in particular, the decorative compass roses, make these artifacts works of art. The first recorded reference to a "portolan chart"

The castle of Bellver near Palma in Majorca, a summer retreat built in the fourteenth century for the kings of Majorca. This Mediterranean island east of Spain was an important crossroads in medieval times. It had been occupied by Moors from North Africa and then conquered in 1232 by an expeditionary force under King Jaime of Barcelona. The Jewish community was vital to the island's wealth and produced the finest mapmakers, among them Abraham Cresques, author of the beautiful Catalan Atlas.

relates to the Genoese ship which carried the crusading French King Louis IX from Aigues-Mortes to Tunis in 1270. The earliest surviving example is the Carte Pisane, which was sold by a Pisan family in the nineteenth century. However, its script is considered by experts to be Genoese. From these fragments of evidence it is fair to conclude that the Genoese were the first producers and users of this revolutionary new form of map.

Certainly the Vivaldos would have sailed with a sea chart to guide them across the Mediterranean and through the Strait of Gibraltar. Within the Mediterranean the early sea charts were sure guides for navigation. They took no account of the curvature of the earth and lacked any form of Ptolemaic grid, but in such a relatively small sea this hardly mattered. They were simple to use. Rhumb lines or loxodromes stretched out from compass roses in all directions, creating a series of criss-crossing lines. The navigator selected a course which ran in his desired direction and reading from a compass placed on the chart he could be sure that the boat was always pointing the right way. The distance he travelled each day was calculated by "dead reckoning," in other words an informed guess. The key to dead reckoning was the boat's speed. One way to ascertain this was to throw a log of wood overboard (hence the origin of a ship's "log"), and measure with a sandglass how long it took to pass from below a crew member at the bow to another standing at the stern. Experienced captains were probably just as happy to guess their speed by intuition and experience.

The vagaries of sail and wind did not usually allow progress directly towards a destination. How therefore could true progress be calculated? The new technology of compass and chart provided an answer. The seaman timed with a sandglass how far the ship had travelled on a particular tack. The line representing this tack formed the hypotenuse of a right-angled triangle. When this line was joined to the compass

A page from the Catalan Atlas, *which was made around 1375 by Abraham Cresques as a gift for the King of France. This spectacular map portrayed the knowledge and temptations beckoning Europeans on the eve of discovery. Most tantalizing of all were stories about the Mansa of Mali, the fabulously rich West African ruler whose wealth was founded on gold. The quest for gold inspired the first Portuguese voyages along the African coast.*

course by a line forming the third side of the triangle the progress achieved along the direct course could be calculated. It was then a relatively simple mathematical step to work out how far it was necessary to sail on the opposite tack to rejoin the direct course. In fact sailors did not need to make these calculations as they had all been worked out in advance on traverse tables, which supplied the answer to every possible variation of distance and direction. This form of navigation was splendid in the well-charted waters of the Mediterranean and Western Europe. The problem for the Vivaldos, however, was that charts followed rather than preceded exploration. Once they turned south into the Atlantic they were in uncharted waters. However far the expedition reached, they could never have imagined the true scale of the target they had set themselves. As the Portuguese would later discover, Africa seemed endless.

Not surprisingly, the expedition vanished. Thirty years later the Genoese Annals speak of the disappearance of the adventurers and the total absence of all news of them. Precisely why and where they disappeared remained a mystery. The Italian name of Allegranza island in the Canaries, just north of Lanzarote, suggests that at the very least the Vivaldos put in there. Indeed they may have been the first Europeans to rediscover these islands which, under the earlier name of the Fortunate Isles, had been the westernmost tip and prime meridian of Greek and Roman maps. A hundred and sixty years later a Genoese sailor in a Portuguese fleet reported that he talked to a man in Gambia who claimed to be the last survivor of the expedition! For over two centuries there were reported sightings of survivors all over the coast of Africa, but no clear facts emerged. For nearly two hundred years after the Vivaldos' voyage the impetus for such long-range discovery was lost. The uprooting of the Mongol empire in the mid-fourteenth century closed the land gateways to the East. The Black Death and intellectual stagnation of late-medieval Europe limited global ambitions. Sea charts, the art of which was still dominated by the Genoese and in particular such practitioners as Petrus Vesconte, concentrated on ever-greater precision within the Mediterranean and along the coasts of Western Europe. Progress was piecemeal and the breadth of vision which had propelled the Vivaldos was replaced by new trading opportunities nearer to home. In the early fourteenth century a Genoese expedition under Lanzorotto Malocello rediscovered the Canaries, their leader giving his name to one of them. The exact date and circumstances of this venture are unclear; the only certainty is that the islands are clearly marked on a chart for the first time in 1339.

A second important center of chartmaking grew up on the island of Majorca, which in medieval times had great strategic and political importance. It had been occupied by Moors from North Africa but in 1232 it was conquered by an expeditionary force under King Jaime of Barcelona. In the decisive battle in the bay of Palma, 50,000 people were said to have been slaughtered. But Majorca remained a cosmopolitan place; settlers from Spain, France and Genoa assimilated much of the Moorish population, who were entitled to buy indemnity from enslavement or expulsion. The most multi-talented people were the Jews, who were described around the time of King Jaime's victory as a "treasure house . . . from whom the trades and traders of this kingdom in peacetime derive great abundance." For the first hundred years of Barcelonan rule, despite some restrictions imposed on their rights to foreign trade, the Jews were more than tolerated. In return for their status as "the King's coffers of money" Jaime decreed that no Jew could be convicted by Christian or Moorish testimony.

The Jews of Barcelona were the most expert mapmakers of their time, and the most distinguished was Abraham Cresques. By his time, the latter half of the fourteenth century, toleration of the Jews had markedly decreased, sparked by the indigenous Europeans' historic paranoia of being overrun. However, Abraham's great talent and his post as cartographer to the king of Aragon entitled him to exemption from the usual restrictions. He was not forced to wear the dress prescribed by the Lateran Council; he enjoyed the rare privilege of running water in his home; and butchers in Palma were instructed by royal command to provide him with special meat so that he might create the richest possible color in his illustrations. Abraham's masterpiece was the *Catalan Atlas* of 1375. It was a gift from the King of Aragon to King Charles V of France and survives in magnificent condition in the Bibliothèque Nationale in Paris. It is a luscious, intricate vision of its time, adorned with exquisite vignettes of kings and palaces, camel caravans and cities, islands and ships, highlighted with gold inlay and rich browns and greens. It contains illustrations of such classical legends as that of the pearl-divers in the Indian Ocean who as if by magic were protected from sharks. Then there were the diamond-seekers who threw meat into the mountain clefts; the diamonds stuck to the meat, which was then brought down by birds.

When describing those areas with which Cresques was more familiar, the *Catalan Atlas* offered a definitive and enticing document for the daring entrepreneur. The most mouthwatering prospect concerned the legend of the Mansa of Mali, the ruler of a fabulously wealthy kingdom in the heart of Africa: "So abundant is the gold which is found in his country that this lord is the richest and noblest king in all the land." The problem for Europeans was the precise location of this huge source of gold. Not surprisingly the Mansa of Mali and the merchants of the Saharan caravans through whose hands the gold was traded had every intention of keeping this strictly secret.

"The gold generated bizarre theories about its origin," writes the historian Felipe Fernández-Armesto. "It grew like carrots; it was brought up by ants in the form of nuggets; it was mined by naked men who lived in holes."

The *Catalan Atlas* included the story, with an accompanying illustration of his boat, of the Majorcan explorer Jaime Ferrer, who in 1346 had gone in search of a rumored River of Gold which poured its wealth into the Atlantic. Like that of the Vivaldo brothers, Ferrer's ship disappeared, but the idea of the River of Gold persisted and was highlighted by Abraham. More than anything else, it was the lure of African gold which drove the next wave of exploration down the coast of that continent. Racked by internal conflicts, the Genoese lacked the will for a concerted drive into the new lands, and the mantle passed to the Portuguese.

"Oh, thou prince little less than divine! Thy glory, thy praises, thy fame, so fill my ears and employ my eyes that I know not well where to begin. Know that thou wilt not find another that can equal the excellency of the fame of this man. With reason mayst thou call him a temple of all the virtues. The seas and lands are full of

Camel trains unloading their eastern cargo on the shores of the Mediterranean. The overland trade from China and the East had made the fortunes of the Italian merchant states. But if the Portuguese could find a sea route to the East, they could cut out the middlemen and even corner the market. For this reason Venice and the other states of northern Italy were anxious to keep abreast of Portuguese navigational achievements, sending spies to Lisbon to steal or copy maps.

Henry the Navigator, Prince of Portugal. His soubriquet was bestowed by an admiring English biographer in the late nineteenth century, but in fact Henry made only two short sea crossings, and his father and brother were as influential in instigating the Portuguese discoveries. Henry's sailors were lured into the unknown by the prospect of slaves and gold.

your praises for that you, by numberless voyages, have joined the East to the West." Thus did his biographer, rather hagiographer, Gomes Eannes de Azurara, panegyrize Henry the Navigator of Portugal, though the soubriquet of navigator was bestowed on Prince Henry by his nineteenth-century English biographer. In fact, Henry only ever made two short sea crossings, the first for the Portuguese invasion in 1415 of the Moorish city of Ceuta where he first became bewitched by African gold and the exotic products of the East which were transported overland from ports on the Indian Ocean.

Whatever the facts of his voyages, Prince Henry became the symbol of the remarkable Portuguese exploration of Africa which, in a piecemeal and unplanned way, would lead to Vasco da Gama's voyage around the Cape of Good Hope to India. Medieval Portugal provided a marked contrast with Genoa: poor peasants eking out a living from a land half covered with stony, scrubby mountains; small towns; few luxuries; even fewer merchants and traders. By the end of the fourteenth century, however, this sleepy nation would be galvanized by two vital and interactive forces. One had been there since eternity—the sea. The other was a strong, centralized monarchy. In 1385 the Aviz dynasty, after defeating with the aid of English archers the King of Castile at the battle of Aljubarrota, assumed the throne of Portugal. King John I and his three sons, the youngest of them Henry, set about building a maritime empire.

Henry, by dint of commissioning a sycophantic biographer, has taken history's laurels. This was particularly unfair on his brother, Dom Pedro, who probably took a greater interest in navigational problems and the broader world picture. It is known that he toured Europe, in particular Italy, in search of the latest maps and information and it was probably owing to him that the legend arose of Henry's salon of savants, navigators and astronomers at Sagres busily pursuing the latest geographic theories of the times. In fact, Henry's headquarters was down the Portuguese coast at Lagos and, if there was such a pioneering establishment at Sagres, there is little to connect it with Henry. After being later caught up in a botched coup, Dom Pedro died a reviled rebel; not for him a fawning testimonial.

Azurara's account gave a variety of reasons for Henry's expansionist plans. One was to find Prester John, who had now been relocated to Africa after failing to show up in Asia. It was a simple matter to mingle his identity with that of the Christian king of Abyssinia. Another reason, according to Azurara, was that Henry "keenly enjoyed the labor of arms, especially against the enemies of the Holy Faith." The religious motive for conquest was a vital piece of public relations, as the Pope's approval was required to establish legitimate sovereignty over any new settlements. Henry became Commander of the Order of the Knights of Christ (Knights Templars) under whose emblem, the famous red cross, the Portuguese voyages took place. The great Abbey and Rotunda in Tomar are the Knights Templars' lasting memorials. Chivalry, the disinterested search for knowledge, even religion, were a front, for really Henry and his cohorts wanted to make money. Henry drank very little, but his colonization of Madeira allowed him to do well out of the local wine. He established the local sugar industry and imported wheat, wax-honey and wood. But the ultimate target remained that elusive, sparkling gold. The "sea of sand," wrote a squire in Henry's household, was unnavigable by Christians. Only the "ship of the desert"— the camel—could cross it. So Henry was forced to find a way round in seagoing ships.

Slowly the Portuguese caravels ventured on, to Cape Noun, Cape Juby and Cape Bojador, where no man had dared to tread. Such at least was Azurara's view of the latter as he regaled his readers with the tales of derring-do by Henry's sailors, ignoring the Phoenicians and indeed more recent Genoese explorations. Clearly though, and with reason, the old tales about the Atlantic cast their spell on sailors emerging from a medieval and superstitious world. Lashing seas, swirling winds whistling around shallow sandbanks, and deadly rocks formed a manifold source of treachery for these early pioneers.

Surprisingly, given that the Portuguese later became so outstanding in chartmaking, hardly any Portuguese charts survive to record their makers' progress in the fifteenth century. There are stories that Abraham Cresques' son, Jafudo, after being intimidated into leaving Majorca by the increasingly oppressive Spanish, went to work for Henry. We know of Dom Pedro's interest in maps. Some Portuguese historians have construed the absence of charts as evidence of all sorts of secret Portuguese discoveries, including that of America before Columbus. But a negative does not prove a positive and whether it was because they were lost in the Lisbon earthquake of 1755 or because they were too undistinguished to survive, the fact is that the visual records of their achievements must be sought in maps made by Italians such as Andrea Bianco and Grazioso Benincasa rather than by the Portuguese themselves.

In these early days exploration was cosmopolitan. The Venetian trader, Alvise da Cadamosto, sailed on a Portuguese expedition in 1455. His famous account of the journey in the tiny 60 ft (18 m) caravels describes the regular pattern of these trips: southwest from Cape St Vincent to Madeira; then on to the Canary Islands to restock; then a tack back to the southeast before edging ever further along the coast of Africa. Cadamosto reached the Senegal and Gambia rivers where he described a hostile encounter with African canoeists with whom there was a sharp exchange of arrows. "In short space a great number of negroes were wounded. By the grace of God, however, not one of the Christians were hit. We then attempted to parley with the negroes. They replied that they did not want our friendship on any terms for they firmly believed that we Christians ate human flesh, and that we only bought negroes to eat." If not in these exact terms, the African fears were nevertheless justified, for these were the early days of the slave trade. Cadamosto wrote that already 1000 slaves were being sold in Lagos every year, forerunners of the millions later sent to work and die in the New World. Now slaves, gold and the abuse of religion became the unholy trinity which drove the Portuguese on.

During his second voyage Cadamosto claimed the first sighting of the Cape Verde islands, but an even more important discovery followed. The crew began to lose sight of the Pole Star to the north and, turning their heads, they made the first recorded sighting of the Southern Cross whose shape Cadamosto drew in his journal. They were still a few degrees short of the equator, but the southern hemisphere beckoned. The torrid zone had not toasted them to death nor had the Ocean Sea enveloped them in an embrace of impenetrable fog. In 1457 the Portuguese, anxious to know where all this was leading, commissioned a revised world map from the greatest expert in Europe. His name was Fra Mauro and he lived and worked in the monastery of San Michele island off Venice. Today the island is the city's cemetery and much of the monastery no longer exists. But there is still a library—and a librarian, an eighty-five-year-old monk called Padre Vittorino whose imagination is still captured by his illustrious predecessor who pulled together everything that was known about the world more than five hundred years ago. Fra Mauro's original map was lost but two years later he was commissioned by the Doge of Venice to make a copy. According to Padre Vittorino, a Venetian senator visited Fra Mauro's workshop to view the as yet unfinished map. He tried to find Venice but could not.

The monastery of San Michele in Venice, which in 1457 was home to the monk Fra Mauro, one of the foremost geographers of his day. Commissioned by the King of Portugal to make a world map with the latest available knowledge, Fra Mauro challenged Ptolemy's belief that it was impossible to sail round Africa.

When Fra Mauro pointed to a dot which seemed tiny compared with the rest of the world, the senator recoiled in amazement and horror. "Why so small?" he complained, "Venice should be bigger and the rest of the world smaller!"

The senator's displeasure was understandable. In the fifteenth century Venice was the queen of Europe, her wealth founded on a network of trading arrangements with the world's most powerful civilizations. She was the hub of a wheel whose spokes reached out to Constantinople, Greece, Egypt, Persia and China. After two hundred years of bitter and hard-fought rivalry with Genoa, Venice now dominated the ports of the Levant and Alexandria. In return for providing material support and naval transport for the Crusades and other papal military adventures, she had a special dispensation from the Vatican to barter with the infidel. She dealt not in everyday goods such as timber and food, but in the costly and profitable luxuries of the East. These included the spices, silk and carpets which had wended their way across Asia in a series of complex transactions before being bought up by Venetian merchants, who took the biggest cut of all.

Portuguese caravels were only about 60 ft (18 m) long, tiny ships in the face of hazardous journeys into the unknown. Navigators hugged the coast as far as possible, stopping en route to pick up supplies and fresh water. By da Gama's time, the Portuguese could find their latitude with accuracy, but longitude was still a matter of guesswork.

The closing of the gateway to the East had helped Venice. She was well served by the fifteenth century's status quo and had everything to lose from rival nations forging a direct link with the primary producers of the Far East. But even in 1457 such a prospect must still have seemed no more than a flight of fancy. The Venetians were happy to indulge the Portuguese and give them the benefit of their scholarship. On his map Fra Mauro professed himself convinced that there was a sea passage going around the south of Africa even if this meant contradicting the newly rediscovered work of Ptolemy. The Doge encouraged the King of Portugal to pursue his voyages, perhaps because he felt that one day Venice might become the middleman for African gold. In fits and starts, depending on the commitment of succeeding monarchs, the Portuguese pushed on. After Henry's death they finally found gold at Mina on what became called the Gold Coast. It was not quite Abraham Cresques' bounteous River of Gold but enough to spur further exploration. As the caravels headed east along the Bight of Africa, they must have begun to think that this was the southern coast of a square continent. But when the coast took a sharp turn southwards just above the equator, disillusionment set in.

The architect of the final push was the vigorous and very competent King John II. Before succeeding to the throne he had been put in charge of African policy, and instilled a new nationalist ambition. Only twenty years after the full and free exchange of information with Venice the Portuguese adopted a policy of xenophobic secrecy to protect their access to African gold and slaves. In 1479, after a brutal sea war against its resurgent Spanish rivals, Portugal decreed that any crew of foreign ships found in its zone of navigation should be thrown into the sea to drown. This lethal threat was reinforced by a demand from the Cortes in Lisbon that new lands should not even be recorded on maps, to avoid the risk of leaks. Nautical writings and charts became state secrets and navigators were forced to swear an oath of silence. King John also sought radical improvements in navigation and position finding. He convened a commission of mathematicians to find the best way to take solar observations. The result was the *Rule of the Sun*, a book which showed how to measure precisely the sun's height with an astrolabe. In order to catch the sun when it was exactly overhead it was necessary to take a series of readings during its approach and retreat from the meridian. The maximum reading therefore must be the correct one. All of this information was summarized in the *Regimento do astrolabio e do quadrante*, the first national manual of navigation.

Right Vasco da Gama led the Portuguese expedition of 1497 which finally sailed round the tip of Africa, up the east coast and on to India. His arrival did not signal the discovery of a new world, but linked two old worlds, each with its own maritime system.

Right Belem Tower on the River Tagus in Lisbon, the starting point for da Gama's voyage. The enterprise was backed by foreign capital and employed the services of foreign sailors, so it proved impossible for the Portuguese to keep their discoveries secret. After da Gama's voyage the sale of charts to foreigners became a lucrative if illegal sideline for mapmakers.

In 1482 Diogo Cao embarked on a two-year voyage during which he discovered the mouth of the River Congo. *Padroes*, or stone columns, were placed on new territory to claim sovereignty. In 1487 Bartolomeu Dias sailed beyond the tip of Africa without realizing he had done so. But then a curious decade of stagnation followed. A 1489 map by the German, Henricius Martellus, shows how definite a shape the tip of Africa had assumed in the minds of educated Europeans. Yet the Portuguese, not realizing quite how close they were and stunned by Columbus' voyages west under the Spanish flag, hesitated at the final fence. King John's drive faltered. In 1495 he died; the sea passage to the Indies had eluded him.

Finally in 1497 under the new king, Manoel I, Vasco da Gama set out, amid loud fanfares and confident expectations, from the tower of Belem on the Tagus for his fully planned and well-drilled assault on the target which had eluded and cost the lives of so many European sailors since the time of the Phoenicians. On 20 May 1498 he reached Calicut in India. The key to his success was the realization that any course hugging the coast of Africa was slow, dangerous and difficult. Instead he struck out boldly to the west, sailing deep into the Atlantic and allowing the prevailing winds to drive him around Africa, to give a wide berth to the dangers of the Cape of Good Hope. Da Gama's arrival in India did not, of course, mean the discovery of a new world; it was the linking of two old worlds. Indeed, for his passage across the Indian Ocean, he had required the expertise of an Arab pilot whom he had hired in East Africa. The Indian Ocean was its own maritime world, buzzing with the traffic of Arab and Indian vessels, a long-established network of trade routes. It was particularly galling for da Gama to find this sophisticated system to be dominated by Moslems when one of the aims of Portuguese discovery had been to locate the Christian kingdom in East Africa under the magnificent sovereignty of Prester John, or, presumably by now, his heirs.

Nevertheless this epic voyage was inducement enough for King Manoel to begin celebrating back in Lisbon. Work started on the spectacular Hieronymite monastery on the shores of the Tagus to commemorate da Gama's achievement. With a certain lack of modesty Manoel proclaimed himself Lord of Guinea and of the Conquests, Navigation and Commerce of Ethiopia, Arabia, Persia and India. He wrote to King Ferdinand and Queen Isabella of Spain, who were still wondering whether Columbus' discoveries to the west would turn up anything more than dancing girls and palm trees, a gloating account of the Portuguese finds: great cities and buildings, fine stones of all sorts, mines of golds. Manoel could not resist pressing home the Portuguese advantage: "We hope with the help of God that the great trade which now enriches the Moors of those parts ... shall, in consequence of our ordinances, be diverted to the natives and ships of our own kingdom, so that henceforth all Christendom in this part of Europe shall be able to provide itself with these spices and precious stones." The implied threat to the Moors would soon turn into a brutal and bloodthirsty religious oppression as the Portuguese expanded their empire and forced Christianity down the throats of their new subjects.

News of the Portuguese breakthrough filtered through to northern Italy. Precise information about their new route and the trading prospects it promised could be found only on maps. In 1502 a shadowy individual called Albert Cantino arrived in Lisbon and set up as a horse dealer. This was a cover, for in reality he was a spy of Ercole d'Este, Duke of Ferrara and one of Renaissance Europe's grandest men. The great castle in the center of Ferrara housed a salon for the most distinguished artists and scientists of the day. From it radiated a planned city of straight streets, of which the most splendid, called the Corso Ercole after the Duke himself, was lined with

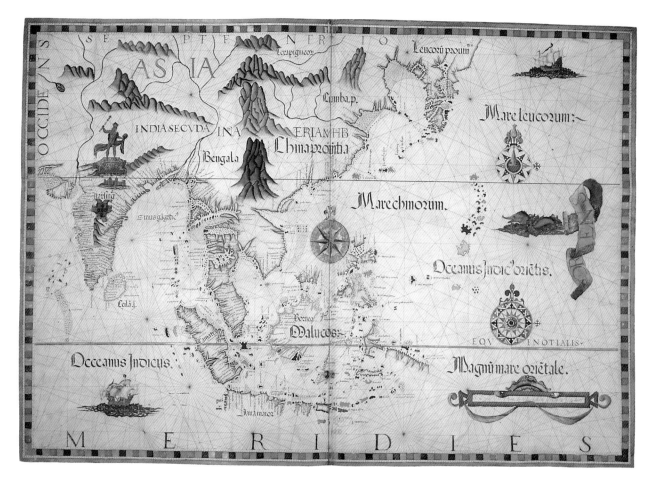

Diogo Homem's map of Indonesia and the Spice Islands provided an early detailed picture of the region, and contained information much sought after by rival nations. When Homem fled Lisbon accused of murder, he sold his expertise to the English and Italians.

sumptuous palaces. Whether out of simple curiosity or because he detected a threat to his wealth, Ercole pressed Cantino for an up-to-the-minute report from Lisbon.

Somehow, probably by bribery, Cantino obtained in 1502 a copy of the latest chart. In this meticulously illustrated map the world as we know it is finally beginning to take shape. It shows the new outline of Brazil, recently discovered by Pedro Cabral, illuminated with three red parrots and a legend that reads: "This land is believed to be a continent in which there are many people who go about naked as their mothers delivered them." The voyage of the Corte Reals to Greenland is marked and the land claimed for the King of Portugal. The Caribbean islands of Columbus are acknowledged as the province of the King of Castile. But laid before inquisitive Italian eyes was the alarming and crystal-clear vision of the direct passage around southern Africa and thence to India and the Far East. Cantino risked his life in procuring such a complete picture. After Cabral's voyage to Brazil another Italian agent had written: "It is impossible to get a chart of this voyage because the King has decreed the death penalty for anyone sending one abroad." Shortly afterwards another copy of a Portuguese world map, made by the Genoese Nicolaus Caverio, also escaped from Lisbon.

Slow at first to realize the threat, the grandees of Venice eventually lumbered into action. The city's ambassador to Egypt enlisted the support of the infidel sultan against the expansionist Portuguese, for both stood to lose by the presence of rivals in the spice trade. In 1504 the Venetian spice council sent their own agent, Lunardo da Ca'Masser, to Lisbon to gather intelligence on the new route to the East and the convoys returning to Lisbon with their cargoes of spices. The Italian historian Corradino Astengo records that after two years and a string of troubles, including a spell in prison, da Ca'Masser returned to Venice with a detailed report on Portuguese navigation. His reward was an attractive sinecure as a university chancellor. Italian interest in Portuguese expertise did not abate. There is a report from 1517 of one Alexander Zorzi who drew up a document containing information on recent voyages and a detailed illustration of the latest Portuguese nautical astrolabe.

The Portuguese empire spread. They built fortress towns at Goa and Calicut in India, and then Malacca in Malaysia. One of Manoel's officials documented the grand design thus: "Whoever is lord of Malacca has his hand on the throat of Venice. As far from Malacca to China, and from China to the Moluccas, and from the Moluccas to Java and Sumatra, all is in our power." In December 1511 three Portuguese ships set sail for the Moluccas of Indonesia, the fabled Spice Islands. It seems strange today that spices could have exerted such a magnetic attraction for European nations, but they were literally vital. Cloves were thought to possess medicinal properties, in particular as a cure for the Black Death which had so ravaged Europe. Furthermore, when rubbed into meat, they acted as a preservative.

These spices, along with nutmeg, mace and cinnamon, grew in abundance in the Moluccas. Aided by local navigators and a Javanese chart, the Portuguese snaked their way between the islands of the Banda and Molucca Seas. Their target was the volcanic island of Ternate, studded with clove trees over 100 ft (30 m) high and monopolized by the sultan's family since time immemorial. The expedition was charted by Francisco Rodrigues and as a result of his hand-drawn illustrations, coupled with existing Javanese charts of these complex waterways, the accurate shape of the Moluccas appeared for the first time on European maps.

Then, just as they arrived to stake their claim, the right of the Portuguese to the Spice Islands trade was suddenly undermined. The discoveries of Columbus had provoked an unseemly clash between the renascent and rapacious thrones of Spain and Portugal. On Columbus' return the Pope granted to Spain the right to all lands in the regions explored by the Genoan explorer. However, under an earlier Papal Bull of 1481 the Portuguese claimed the right over all land and waters south of the Canaries and west of Guinea. Portugal, which had refused to fund the venture in the first place, now had the audacity to claim that Columbus' new lands belonged to her. The Pope, Alexander VI, whose scandalous private life concealed a finely tuned administrative mind, restored matters in favor of his native Spain by decreeing that any lands to the west of an imaginary north-south line drawn 100 leagues (300 miles/480 km) west of the Azores should be reserved for Spain. Such a ruling, at a time when Columbus' discoveries were still thought to be off the coast of China and Japan, would have disastrously curtailed Portuguese ambition.

The dispute was finally resolved between the two squabbling kingdoms at the Treaty of Tordesillas, a small town in northern Spain, in 1494. The line was moved further west. That half of the world which lay to its east would belong to Portugal; the half to the west would belong to Spain. It turned out to be King John II of Portugal's final masterstroke even if he never lived to see its full consequences. The as yet undiscovered land that would become Brazil would fall into Portugal's half and, as Columbus' lands turned out to be two new continents, Portugal could still retain its rights to India. The breathtaking arrogance of the two Catholic nations in simply dividing the world between them was never an issue.

However, the exact dividing line on the other side of the world was another matter. As new maps took shape it became clear that this line might fall somewhere

The island of Ternate in the Moluccas of Indonesia. This was the most important of all the Spice Islands, because it was the source of cloves, believed in the Middle Ages to be a cure for the plague. Cloves were also used for preserving meat, a vital function in the centuries before refrigeration.

The bustling harbor of Lisbon around 1550, at the height of the Portuguese empire. Capture of the spice trade had been a remarkable achievement for a nation which only 200 years before had been an undeveloped land of peasant farmers. However, by the end of the century the Dutch, aided by Portuguese maps, were challenging Portugal's hold on the Spice Islands.

around the treasured Spice Islands which, not surprisingly, the Spanish crown dearly wished to lay its hands on. In 1517 Spain's chance came with the arrival in Seville of Ferdinand Magellan. Magellan had served for years in the Portuguese armed forces: he had fought with the fleet which captured Malacca and may himself have sailed to Ternate and Tidore. Then in a campaign in Morocco he was accused of trading with the enemy. King Manoel refused to clear his name and give him the promotion which Magellan believed that after years of loyal service he deserved. Persuaded by the mathematician and astrologer Ruy Faleiro that the Spice Islands could be reached by a short crossing from South America, Magellan deserted his native country and offered the idea to Spain.

Magellan was not the first renegade. By now Portugal possessed unsurpassed cartographic skills but refused to pay the expert mapmakers properly. The Spanish stepped in smartly and lured some of these disillusioned craftsmen to Seville to work in the Casa de la Contratación, the trading body which had been given the rights to exploit the new lands. When Magellan arrived he found the distinguished chart-maker Diogo Ribeiro already working for the Spanish and passing on his detailed knowledge of the Far East and the Spice Islands. A report from the Portuguese ambassador in Seville to King Manoel relates that another Portuguese mapmaker, Jorge Reinel, had made a map and globe specifically to help Magellan prepare for his voyage. Reinel had got into trouble in Lisbon and fled the country. His father, Pedro

Reinel, also a mapmaker, was one of King Manoel's most trusted servants. Pedro travelled to Seville to bring his son back to Lisbon. Ironically the ambassadors' report then claims that Pedro himself had been seen adding in details of the Spice Islands on the Spanish map; even worse, he drew a line of demarcation placing them in the Spanish half of the world. Somehow Pedro Reinel escaped notice or censure back in Lisbon, as later documents record that he was awarded a generous pension.

In 1519 Magellan set sail. The lethal seas around Cape Horn gave way to the deceptive calm of the Pacific, named after the endless smoothness which drove Magellan's crew to such desperate measures in an attempt to survive. The famous account by Antonio Pigafetta shows how gruelling this supposedly short crossing became: "We entered into the Pacific Sea, where we remained three months and twenty days without taking in provisions or other refreshments, and we ate only old biscuits reduced to powder, and full of grubs, and stinking from the dirt which the rats had made on it when eating the good biscuit, and we drank water that was yellow and stinking. We also ate the ox-hides which were under the main-yard and also the sawdust of wood and rats. The upper and lower gums of most of our men grew so much that they could not eat—and nineteen died." Eventually they reached the island of Guam. A few weeks later Magellan was killed in the Philippines. His crew sailed on, putting in at the island of Tidore, where they were welcomed as possible allies against the oppressive Portuguese, and loaded with a cargo of cloves. Three years after they set out, one ship, the *Victoria*, staggered into Seville, her cargo of cloves still intact. The world had been circumnavigated, and it was bigger than anyone since Eratosthenes had imagined.

The quarrel between Spain and Portugal over the ownership of the Spice Islands dragged on. The archives record an astonishing meeting in the border town of Tomar between the Spanish Ambassador to Portugal and the Portuguese Diogo Lopes de Sequira, a former Governor of India. The disgruntled Portuguese made a remarkable offer to secure a copy of the crucial charts for the King of Spain. The Spanish Ambassador reported that "the one who makes those [charts] of the King of Portugal is called Lopo Homem with this one negro assisting; they live in Lisbon and have orders to make no chart for anybody but the King. But sometimes they venture at a price though it is very difficult." The Lopo Homem in question was master of all navigational charts in Portugal, the king's chief of naval intelligence. Yet he was offering to sell the state's greatest secrets. The hemorrhage of vital information continued. In the 1550s Lopo Homem's son Diogo fled the country after killing a man in a brawl in the notoriously violent quarter of Lisbon where the mapmakers gathered. The king was so anxious not to lose him and his secrets that he promised to waive all charges if Diogo Homem returned to Lisbon. But he stayed away, selling his skills and knowledge first to the English and then to the Venetians. In 1558 he presented a famous atlas to Queen Mary which is today in the British Library.

Knowledge of the seas and accurate charts proved vital factors in creating the physical and political shape of the world. The thrust of exploration had been, either by the East or by the West, directed at the riches of the East. On the way Columbus discovered new lands which would provoke intense curiosity and a geographic craze in Europe. The mapping of these lands would also raise a quite different issue: by what right did one man or one nation claim possession of a territory?

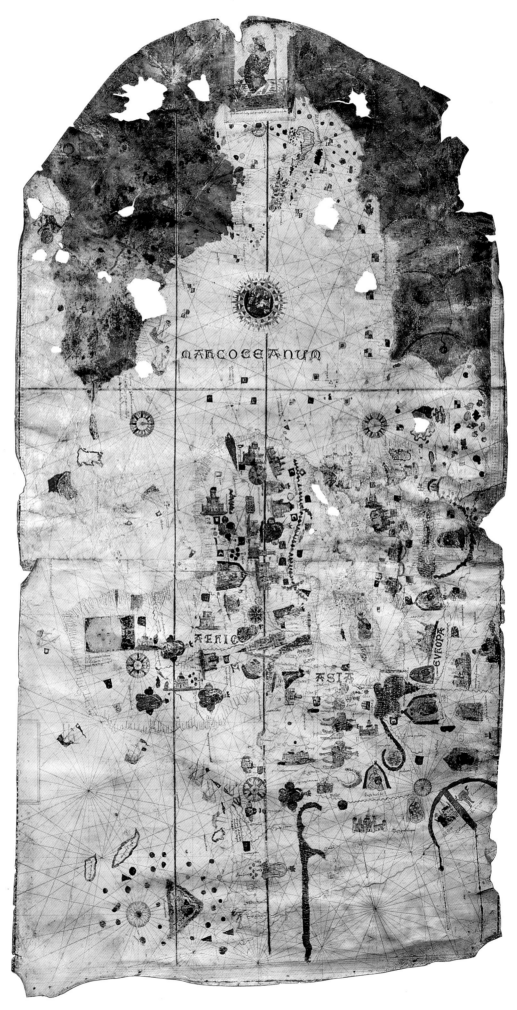

MARCOCEANUM

AFRIC

ASIA

EUROPA

The first map to show the emerging outlines of the New World, drawn around 1500 by Columbus' pilot, Juan de la Cosa. Its overall form is taken from the sheep's hide on which it is drawn; west is at the top, with an image of St Christopher. When the map is turned to the left, the shapes become more familiar. The bulge of Africa is in the middle; in the sea to the center left are the Caribbean islands; and above and below, de la Cosa's best guess at the North and South American coastlines.

5

OLD WORLDS, NEW WORLDS

America's birth certificate is to be found in an unexpected place. In the south of Germany, hundreds of miles from the Iberian and English ports from which the first European explorers set sail, lies the small Bavarian town of Wolfegg, dominated by a castle which is the ancestral home of the Waldburg-Wolfegg family. Prince Max Willibald Waldburg-Wolfegg is the proud owner of the only original copy of a world map made in 1507. The map, which was rediscovered by chance by a Jesuit priest working in the castle at the turn of the century, is the first to christen the New World America. It is a potent symbol of the gigantic act of imposition by which the nations of Europe dispossessed the indigenous peoples of their land and shaped it in their own image.

America was named after the Florentine businessman Amerigo Vespucci. In a late-sixteenth-century engraving Vespucci, having just landed on the coast, is depicted encountering America as represented by the attractive and buxom figure of an Indian maiden. It is not hard to infer from this scene the European view of a new virginal land ripe to be seized and raped. It was a historical fluke that the land was named after the Florentine. Although he claimed to have been on four voyages to the New World, Vespucci probably went on just two, the first in 1499 under the Spanish flag and the second in 1501 with the Portuguese. He was a fluent writer and a polished diplomat whose letters earned him such a reputation as an expert geographer that in 1508 he was appointed Pilot Major of Spain, where he remained until his death in 1512. Vespucci claimed that he, not Columbus, had first set foot on the main landmass of the New World as distinct from the islands of the Caribbean and that he first realized that it was a new, fourth continent. While the latter may have been true, the former certainly was not, but Vespucci's account prevailed in the academic circles of Europe. Nowhere were his writings more voraciously read than in the German town of Saint-Dié (now in eastern France) in the Vosges Mountains, where an elite group of thinkers including the Canon of Saint-Dié, Martin Waldseemüller, formed the Gymnasium Vosgense. This learned circle devoted itself to philosophy, cosmography and mapmaking.

In many respects the scholars of Saint-Dié were traditional followers and interpreters of Ptolemy. But although the new discoveries did not figure in

Christopher Columbus, whose attempt to find a trade route to China and the Spice Islands by sailing west rather than east led him to discover the New World. He never fully realized the significance of his find, assuming at first that he had hit on some islands off the coast of Asia. His seizure of the new lands from their inhabitants started the process which would all but destroy the native populations.

Ptolemaic ideas, the learned circle had imbibed the news of them and were eager to place them in a theoretically correct framework. In 1507 Waldseemüller, working from a copy of the latest Portuguese map made by the Genoese Nicolaus Caverio, which had been spirited out of Lisbon two years earlier, decided that his great map should honour Vespucci above all. It was an error he lived to regret and he later tried to call the world's attention to his realization that the honors should instead go to Columbus. But it was too late: his nomenclature was by now irrevocably imprinted on the European consciousness. It could be argued that Columbus brought this travesty of history upon himself because he never understood the true significance of his discovery. An avid reader of the cosmographies of his time, Columbus hoped to find in them a practical scheme which would enable him to sail from Europe to China. First he drew on the works of Cardinal Pierre d'Ailly, whose famous *Imago Mundi* of 1410 was the last great summation of academic medieval geography. For Columbus, what was most significant about d'Ailly's work was his recognition that the east-west extent of Asia had been overestimated. This meant that it would not be necessary to sail halfway round the world to reach China as Ptolemy had suggested.

Another important influence was Paolo dal Pozzo Toscanelli, a physician and member of a prominent Florentine family of merchant bankers with whom Columbus exchanged letters around 1480. Toscanelli had studied the travels of Marco Polo and other travellers to the East and had previously traded this information with Portuguese visitors to Italy, who offered him in return news of the southward progress of the Portuguese in Africa. In 1474 his Portuguese friends, daunted by the apparently interminable length of the west coast of Africa, informally consulted Toscanelli about the possibility of short-cuts. Toscanelli replied that China was only 5000 miles (8050 km) to the west and could easily be reached by an enterprising navigator as there were plenty of islands on the way.

Islands in the Atlantic or Ocean Seas were a common creation of both classical and Christian writers, who usually drew on legendary voyages. For example there was St Brendan's Island, which from the ninth century became known as Brasil. This island had no connection with the South American nation of Brazil, which was named after the red brazilwood which grew there, but was derived from the Gaelic *breas-ail*, meaning blessed. Further south there was Antilia, the Isle of Seven Cities, which was seen by Iberian explorers as a crucial stepping stone on the route west. According to this tradition, seven Spanish bishops had escaped the Moorish invasion by sailing west with their congregations and settling on this island where their descendants could still be expected to live. Substance was given to the idea of these legendary islands by the existence of the four main groups in the mid-Atlantic: the Azores, the Madeira group, the Canaries and the Cape Verde Islands, all of which had by the time of Columbus been discovered or found afresh.

It was not enough for Columbus' purposes that the extent of Asia had been overestimated and that there were stepping stones in the Atlantic. He required an even more drastic reduction of the size of the world to shorten his journey. In d'Ailly's *Imago Mundi* he found the answer. D'Ailly had used the Arab cosmographer Alfraganus as his source for the calculation of a degree of latitude whch equalled a degree of longitude at the equator. Alfraganus settled on the figure of $56\frac{2}{3}$ miles but, like the Greek stade of antiquity, the mile could have many different lengths. Naturally Columbus chose the most convenient interpretation and assumed that Alfraganus was using the short Italian mile. This reduced the size of a degree to 45 nautical miles (83 km) and the journey from the Canary Islands to Japan, seen as the gateway to China on the western route, to 2400 nautical miles (4440 km). In fact the distance is over 10,000 nautical miles (18,520 km). Columbus' delusion was reinforced when his first sighting of land across the Atlantic turned out to be close to the position he had mapped out for Japan.

Such theories became all the more attractive when displayed on a globe. It was Martin Behaim, a trader from Nuremberg, who made the earliest terrestrial globe to survive. Like Saint-Dié, Nuremberg was an important academic and scientific center. Behaim added practical knowledge to theory during a spell in Lisbon at the height of the final Portuguese push down Africa during the late 1480s. On his return to Nuremberg the city council commissioned him to make a globe that included the latest Portuguese discoveries. However, the most interesting feature of Behaim's globe was not the coast of Africa but the clarity it gave to the idea that only a relatively short sea journey separated Europe from Asia. With the island stepping stones such as Antilia scattered along the way the possibilities were obvious.

Early interpretations of the discoveries were cautious. The first voyages encountered only islands, although Cuba seemed to be larger and was identified by Columbus as part of the Asian mainland. In 1498 on his third voyage Columbus discovered the point of Paria on the continent of South America and further exploration confirmed a landmass stretching into the distance. Columbus now had to admit that what he saw with his own eyes must modify his previous theories: "I have come to believe that this is a mighty continent which was hitherto unknown. I am greatly supported in this view by reason of this great river, and by this sea which is fresh, and I am also supported by the statement of Esdras in his fourth book, the sixth chapter, which says that six parts of the world consist of dry land and one part of water. . . . Moreover I am supported by the statements of several cannibal Indians whom I captured on other occasions, who declared that there was mainland to the west of them." Despite this revision, Columbus could still maintain that even though there might be a new continent to the south, the lands to the north remained the outposts of Asia. If this were so and the Earth remained the size which Columbus had calculated, it might indeed be six-sevenths land and only one-seventh water. The final rejection of this idea had to wait until Magellan's voyage around the world and the discovery of the true extent of the Pacific.

The first map of the New World to survive was made by Columbus' pilot, Juan de la Cosa, in 1500, eight years after the first expedition had set out. The Caribbean islands are shown in detail and the northeast coast of South America begins to assume a recognizable shape. Cuba has become an island but America and Asia are still joined as one continent without an intervening ocean. The first printed map to incorporate the New World, the Contarini map, was made in 1506 and was followed in 1507 by Martin Waldseemüller's woodblock. One thousand copies of Waldseemüller's map were printed—a large edition for those early days of printing—and it dictated the shape of America for the next three decades.

Like the Portuguese, Columbus set out with the intention of discovering a trading route from which new wealth would accrue to his royal sponsors from direct access to the profitable luxuries of the East. His encounter with a new and unsuspected land which lacked a familiar civilization gave rise to a different set of ambitions and temptations. Attitudes towards the native peoples veered from the idealized vision of the "noble savage" when they were friendly, co-operative and submissive, to the desire to banish them from the human race when they decided to resist the invaders. Rapidly the latter view prevailed and there was ruthless exploitation of the indigenous populations by Spanish settlers. The statistics, as recounted by Tzvetan

Columbus' ship, the Santa Maria, *runs aground in Haiti. His voyage was based on a massive underestimate of the Earth's circumference, for by selecting evidence from a variety of writers and geographers Columbus had concluded that the journey from the Canary Islands to Japan would be no more than 2400 nautical miles (4440 km). In fact the distance exceeds 10,000 nautical miles (18,520 km)*

Todorov in *The Conquest of America*, say more than any description: "Without going into detail, and merely to give a general idea (even if we do not feel entirely justified in rounding off figures when it is a question of human lives) it will be recalled that in 1500 the world population is approximately 400 million, of whom 80 million inhabit the Americas. By the middle of the sixteenth century, out of these 80 million, there remain ten. Or limiting ourselves to Mexico: on the eve of the conquest, its population is about 25 million; in 1600, it is one million. . . . If the word genocide has ever been applied to a situation with some accuracy, this is here the case." The import of new diseases, slavery, and the loss of the will to live and breed in a land usurped took an unparalleled toll.

Maps, which were the only way in which Europeans could visualize their dominions in the New World, supported this abrogation of the rights of the indigenous peoples. Those who made them for publication were no doubt simply doing their best to make theoretical and scientific sense of the news brought home with every voyage. Yet their notions of sovereignty and territory were unconsciously cast in a blinkered European mold. Columbus' first act on disembarking was to draw up a deed of possession in front of the bemused native onlookers: "He called upon them to bear faith and to witness that he, before all men, was taking possession of the said island— as in fact he then took possession of it—in the name of the King and of the Queen, his Sovereigns."

There followed the naming of the islands—San Salvador, Santa María de Concepción, Fernandina, Isabella, Juana. Columbus was fully aware that all these islands carried names granted by their native inhabitants but the question of preserving such names and recognizing any rights of indigenous ownership simply did not arise. In fact he scattered his names all over the Caribbean: a cape is named Formosa because it is beautiful; another, covered with palm trees, is called Cabo de Palmas. Mapmakers at home took such attitudes for granted. Those such as Waldseemüller had not thought of investigating local names or sovereignty. The interiors were supposedly filled with cannibals, strange animals or impenetrable forests, all of which could be seen as obstacles to be removed. Meanwhile territorial ownership was being unhesitatingly transferred to the invading Europeans.

The unorganized peoples of the Caribbean were in no position to resist this imposition, and their numbers simply dwindled. But what happened when Europeans came into contact with the great civilizations of Central America? In 1513 a small band of survivors from a disastrous Spanish expedition which had set out four years earlier to settle on the Central American coast provided the final confirmation of a new continent. Columbus' belief that he had found the eastern edges of Asia was proved to be an illusion. Under their leader, Vasco Núñez de Balboa, they had settled on the Darien coast. From the local Indians Balboa learned of a further coast to the west, just a few days' march away. With about one hundred men Balboa trudged through the forests of Panama and from a hilltop saw a new ocean disappearing into the distance. It was clear that in order to reach China and the Spice Islands this new sea would have to be crossed.

Balboa's discovery clarified two separate opportunities for adventurers in the service of Spain. The first, the direct eastern route to the East, was demonstrated by Magellan with his voyage around the world in 1519–21. The second was to seize the wealth of the New World. For nearly thirty years after Columbus' voyage this ambition was approached in piecemeal fashion. Settlements on Caribbean islands, explorations up and down the coast of Yucatán and hesitant attempts to establish a foothold on the mainland did not add up to a concerted imperial push.

Right *The Aztec Emperor Montezuma being borne on his carriage to meet Cortés, in a Spanish painting of 1698. Montezuma's empire was oppressive and terrifying, for the Aztec belief that the sun would only rise each morning if nourished by sacrificial human blood demanded the ritual slaughter of prisoners from subject peoples.*

Right *In an ancient custom of the Totonacs – the first subjects of the Aztecs whom Cortés encountered – four Voladores, or Flyers, descend from the top of a 100-ft (30 m) pole. The ritual was a model of the pre-Columbian Mexican universe, the flyers representing north, south, east and west.*

On 22 April 1519, Hernán Cortés landed on the coast of Mexico. His vision would convert Spanish disorganization into the world's greatest empire. He did it without royal authority and in defiance of the orders of the Governor of Cuba, Diego Velázquez, under whose auspices his expedition was despatched. Velázquez was seeking jurisdiction from the Spanish Crown to conquer new lands on his own behalf rather than that of his superiors in the Indies. He instructed Cortés simply to search for a fleet which had gone missing the previous year and to find any Christian captives in Yucatán. Cortés was also entitled to explore and trade but not to colonize or settle. Such rights, even if Velázquez had wished to grant them to Cortés, had to be sanctioned by the Crown.

Cortés ignored these instructions and instead embarked on a gamble for immense stakes. His expedition was small: 550 men, 16 horses, 14 cannon and a few dogs. On landing he had taken possession of the new territory in the manner of Columbus. Bernal Díaz, a soldier who wrote a superb chronicle of the expedition, described how Cortés established his claim: "He did it in this way: he drew his sword, and, as a sign of possession, made three cuts in a large silk-cotton tree which stood in that great courtyard and cried that if any person should raise an objection he would defend the King's right with his sword and shield, which he held in his other hand." But, by contrast with Columbus' approach, this ceremony was designed to impress those of Cortés' soldiers who held allegiance to Velázquez as much as the local Indians. Even so, Cortés had no thoughts of respecting local ownership; his intention was conquest.

Soon after landing, Cortés encountered the Totonac people, whose chief, Tentlil, told him that they were the subject race of a great emperor called Montezuma, the King of the Aztecs. The Totonacs complained bitterly of their oppression by the Aztecs and the tribute which they had to pay to them. It was an easy matter for Cortés to persuade Tentlil that he would free him from the rod of Montezuma if the Totonacs supported him. This early encounter gave Cortés the first hint of a new culture and a quite different view of the world. The Totonac Voladores, or flyers, still perform a spectacular ceremony which has its roots in their ancient cosmology. Amid the ruins of the city of Cempaola five flyers dance their way, blowing their pipes as they go, to the base of a narrow 100-ft (30 m) pole. With no safety net they climb to its summit and then four of them peel away on ropes and spin round the pole, while the fifth remains on his perch at the top beating a drum. The ritual is now a tourist attraction, but nearly five hundred years ago such a dazzling spectacle must have mystified the Spanish interlopers.

The flight of the Voladores represented the world itself. All beings were grouped according to the four cardinal points of the Earth, the flyers representing north, south, east and west. The symbolism extends further as each of the compass points is identified with one of the four creator gods. At the four points could be found trees which held up the sky, echoing the Egyptian pillars which held up the Earth's four corners. Together with the central axis—the pole in the Voladores' dance—the trees formed the connection by which the sacred energies of the gods could descend to earth. One such god was Quetzalcoatl whom the Aztecs prophesied would return to rule over them in the fifth creation of the world, which was now due. This prophecy explains the dread with which Montezuma heard the news of Cortés' arrival and why he was so quick to send him emissaries and offer him gifts. He believed that Cortés was the incarnation of Quetzalcoatl.

Cortés, escorted by Totonac guides and seeking other allies as he went, continued his march towards Tenochtitlán, the Aztec capital. The guides, frightened of

The Mendoza Codex, an Indian map of the founding of the Aztec capital Tenochtitlán (today's Mexico City) in 1325. The priest who led the Aztec migration from their traditional home Aztlan said that when the sun, represented by an eagle, alighted upon a tiny cactus whose prickly pears were like human hearts, they should rest there and found a city.

encountering Montezuma's army, led the Spaniards on a tortuous route through a magnificent landscape of snow-capped mountains and steaming volcanoes. This was not new or unmapped country but was criss-crossed by a sophisticated network of trade routes which had grown up to support a burgeoning civilization. Tenochtitlán was the epicenter of this civilization and, in the Aztec view, of the world itself. The Aztecs developed a tradition of mapping their journey to this final resting-place. Their wanderings began in the year 1168 when they set out from their legendary home in Aztlan. The priest who led the migrations said that when the sun, represented by an eagle, alighted upon a tiny cactus whose prickly pears were like human hearts, there they should rest and found a city. In 1325 they reached the prophesied site and founded Tenochtitlán, which was to become over the next two hundred years the capital of the greatest empire in central America.

But in Aztec cosmology the city meant much more. In Aztec belief, as in the medieval European idea of the chain of being, the universe had many layers. Above was the celestial level, the thirteen heavens or skies; below, the nine levels through which the dead must pass before reaching the underworld. In the middle was the Earth and at its center the great temple of Tenochtitlán. This building stood where the four directions or rhombuses of north, east, south and west intersected and coincided with the terrestrial level of the Aztec universe, at the very navel of the world. Cortés was confronting another vital aspect of the Aztec world. During their march inland his small army arrived at the town of Xocotlan. In the main square Díaz reported that there stood "many piles of human skulls, so neatly arranged that we could count them, and I reckoned them at more than a hundred thousand." These skulls were the relics of human sacrifice. For the Aztecs this rite was a vital engine of the world itself, for the sun, their leader, had to be nourished every day to continue its nightly struggle with the moon and rise again. Its food was prisoners sacrificed by the Aztecs. The sun was the symbol of good, constantly fighting the gods of darkness, the symbols of evil.

Pessimism dominated the Aztec view of the cosmos. In the end the sun would be defeated and die amid fearful earthquakes. The Earth itself was an ally of death. It was the burial place of humans and the hiding place of the stars. Sometimes it was pictured as a monster, part shark and part alligator, sometimes as a fantastic frog whose mouth had great tusks and whose feet and hands were armed with claws. Two months every year were devoted by the Aztecs to honoring the dead and performing human sacrifice to ensure the flow of fresh blood so vital to regeneration. Mankind would be destroyed and life was transitory. With such deep foreboding, the Emperor Montezuma waited for Cortés to arrive. On 8 November 1519 the Spaniards, who had now conquered and won the allegiance of the Aztecs' subject peoples, arrived in Tenochtitlán. They were taken to the top of the temple to view the extraordinary panorama. Díaz wrote: "We were amazed and said that it was like the enchantments they tell us of in the legend of Amadis, on account of the great towers and temples and buildings arising from the water and all built of masonry."

Cortés himself made repeated references to the fact that the Aztecs were more intelligent than the peoples of other islands. "These people live almost like those in Spain, and in as much harmony and order as there, and considering that they are barbarians and so far from the knowledge of God and cut off from all civilized nations, it is truly remarkable to see what they have achieved in all things." He compared their country with his homeland. For example, the most important towers of Tenochtitlán were higher than the cathedral of Seville, its marketplace larger than that of Salamanca. Yet, however great his admiration, Cortés was driven to destroy

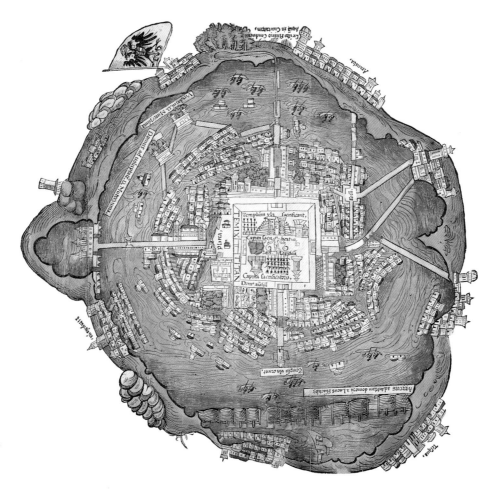

Tenochtitlán as mapped by the Spanish before they destroyed the city. The towers and temples, which were so awe-inspiring that they led some Spaniards to wonder if they were in a dream, can be seen at the center of the city. Cortés reduced them to rubble.

this great civilization. One of his first acts, just six days after he arrived in Tenochtitlán, was to kidnap Montezuma. The safety of his army was one thing but Cortés had an ulterior motive, which was frankly stated in a letter to the King of Spain: "Furthermore, by having him with me, all those other lands which were subject to him would come more swiftly to the recognition and service of Your Majesty, as later happened."

In June 1520, after eight months of uneasy coexistence, the Aztecs finally rose up and the Spaniards were driven from Tenochtitlán. During the uprising Montezuma was killed. Cortés regrouped his army and returned to Tenochtitlán, and by August 1521 the reconquest was over. The once mighty city lay in ruins. The great temple and 78 others in the city's sacred enclosure were razed to the ground. In their place Christian churches were built to impose a new European order over the obliteration of the old. The land was divided up among the victorious soldiers. Cortés' gamble throughout the operation had been to present the King of Spain with a *fait accompli*. Even if he had pursued his conquest without royal approval it would require only the prospect of riches to ensure royal favor. The Aztecs had mapped their empire by means of tribute rolls in which each province was defined, major territories indicated and the amount and type of tribute shown. Cortés was attracted to the gold-producing areas, and with such aids the plunder of America began.

For Philip II, who reigned during the zenith of Spanish power in the latter half of the sixteenth century, after the death of Cortés, there was one obstacle. Transatlantic travel was still a dangerous enterprise and it would have been a foolhardy monarch who deserted his kingdom for the long and risky journey required to inspect his

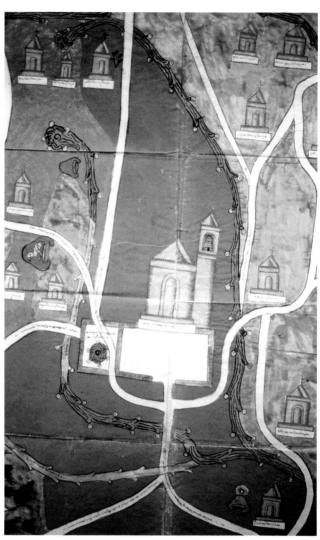

Far Left *Philip II of Spain, ruler in the second half of the sixteenth century of the greatest empire the world had ever known. Transatlantic travel was so dangerous and the king himself so sickly that Philip never saw his vast possessions in the Americas. Instead he commissioned a detailed survey of Mexico called the* Relaciones Geográficas.

Left *A map from the* Relaciones Geográficas, *the survey of Mexico commissioned by Philip II of Spain. Using mainly Indian artists, the survey provided a wealth of information about the geography, population and wealth of each province. Above all it established the spread of Christianity over a society whose own temples and concept of the universe had been obliterated.*

distant dominions. Yet Philip was very curious to know the full extent of his empire, for he was by nature an avid collector. He was also captivated by geography, and the ceiling of the library at his monastic retreat of El Escorial outside Madrid is painted with figures of ancient geographers and astronomers with their instruments. Ptolemy himself is shown hunched over a globe. In 1566 Philip commissioned the mathematician Pedro de Esquivel to make the first detailed map of Spain. One colleague wrote that "there is not an inch of ground in all of Spain that has not been seen, walked over or tramped upon [by Esquivel], checking the accuracy of everything—in so far as mathematical instruments make it possible—with his own hands and eyes." An atlas of twenty-one maps drawn from Esquivel's work survives in the Escorial library today. Philip II's biographer, Geoffrey Parker, states that "the Escorial atlas contains by far the largest European maps of their day to be based on a detailed ground survey. No other major western state of the sixteenth century possessed anything like it."

Philip also ordered the first thorough survey of Mexico, called the *Relaciones Geográficas*. Relying heavily on the work of Indian artists, the survey provided a profusion of information ranging from traditional pictographs to more familiar

El Escorial, the magnificent but sombre monastery which Philip II built as a retreat outside Madrid. Philip was an obsessive collector and numbered among his possessions over 7000 religious relics. He was also fascinated by geography: the ceiling of the library at El Escorial was painted with figures of Greek scientists, including Ptolemy, and its bookshelves were lined with atlases.

maps. Mountains, rivers, churches, towns and buildings were all illustrated, often in attractive colors. By a cruel irony the conquered peoples became the creators of a picture of the world which would satisfy their European masters. Their efforts enabled Philip to survey his newly acquired dominions without going further than his massive library.

The invasion of North America would be a far slower process. Like the Iberian voyages to the south, it was driven from the start by the search for a direct route to the Indies. The Northwest Passage became the Holy Grail which spurred on explorers for hundreds of years and killed many of them, for the climate of the northern latitudes was harsher, the winds more violent. When the first explorers reached land after an Atlantic crossing there were no signs of great civilizations or gold. The Norse discoverers from Norway, Denmark and Sweden had by the beginning of the eleventh century made the crossing to Newfoundland, using their settlements in Greenland as their stepping stone. But two hundred years later these settlements had withered away and exploration had ceased.

In 1497, the same year in which Vasco da Gama sailed around the tip of Africa, the Venetian emigrant John Cabot, under the aegis of King Henry VII of England, sailed west into the Atlantic from the English port of Bristol. In his single ship, the *Matthew*, he reached Newfoundland, discovered its rich fishing banks and returned safely to Bristol. Like Columbus he believed that he had reached an eastern promontory of Asia. In May the next year Cabot set out with a larger expedition. He and his four ships disappeared without trace but for the first time since the Norse voyages a few bold Europeans began to look west across the northern Atlantic.

During the time of these early voyages ideas about North America were far more vague than those about the south. No European nation made a committed attempt to come to terms with the mysteries it posed. There was some enthusiasm in France. François I, who reigned from 1515 to 1547, had reacted haughtily to the Pope's division of the world between Spain and Portugal. "We fail to find this clause in Adam's will," he retorted. François despatched the Florentine seaman Giovanni da Verrazzano to search for the elusive passage to the Spice Islands. Verrazzano's voyages between 1524 and 1528 established the full extent of the unbroken coastline stretching north from Florida to Newfoundland. He concluded without hesitation that this was a new and separate continent. But he also believed that there was a narrow isthmus just north of Florida whch opened out into a great bay leading into the Pacific Ocean. It was a curious error which distorted the shape of America on many maps for the next hundred years.

François I was unimpressed by vague ideas of an empire in the new continent, but he was still keen to find a passage to Asia. In 1534 Jacques Cartier set out on the first of his three voyages. He encountered Indians, and discovered and named the Bay of St Lawrence, yet failed to find any way through. The most attractive report concerned a kingdom called Saguenay which was described by a Huron chief called Donnaconna. Cartier took Donnaconna back to France, where he lived in style for four years regaling the court with enticing stories of the gold and spices to be found in Saguenay. This remarkable place also produced oranges and pomegranates and nurtured a race of winged men who flew like bats from tree to tree. If his Indian informants were sometimes unreliable, Cartier's expeditions nevertheless filled in important pieces of the North American jigsaw. His information stimulated the Dieppe school of chartmakers which flourished in the 1540s and produced manuscript charts of the world rivalled for beauty only by those of the Portuguese. One, known in English as *The Boke of Idrography*, was given by its maker Jean Rotz to Henry VIII of England.

Although Henry VIII was entranced by his new map, the first attempt to create an English settlement in America had to wait another forty years. In the 1580s a small group of ambitious young men gathered at the Middle Temple in London's legal quarter and began to make plans. The search for direct routes to the Indies continued to prove elusive: the Muscovy Company, for example, had failed in several expeditions to find a northeast passage over the top of Asia. They had concluded "that the passage by the Northwest is more commodious for our traffick, than the other by the East." The voyages of Martin Frobisher and John Davis further probed the coastlines of North America but by now the businessmen who backed such voyages were beginning to hedge their bets. All sorts of openings, most notably the Hudson strait, were reported, but all turned out to be dead ends. Explorers were ordered to load their ships with cargoes of cod from Newfoundland to defray the expenses of their expeditions.

The Middle Temple group began to examine the problem from a different perspective. They had all been stirred by the return in 1580 of Sir Francis Drake from his voyage around the world. Drake was an expert seaman, a ruthless leader and a pirate. Not only did he raid Spanish or Portuguese ships for their booty, but he also kidnapped their pilots. After two years and ten months at sea he returned with a cargo of spices and gold which gave his backers a 4700 per cent return on their investment, of which Queen Elizabeth I's share was £160,000. The swashbucklers of the Middle Temple conceived the idea of a permanent settlement in America from which further raids on Spanish and Portuguese shipping could be launched.

In 1584 Sir Walter Raleigh sent an expedition to reconnoiter suitable locations on the North Carolina and Virginia coasts. The commander of one of the vessels, Arthur Barlowe, returned with a full description of an apparently ideal spot. There were grapes, tall cedar trees, deer, rabbits, hare and fowl in incredible abundance. They found some friendly Indians, one of whom they invited onto their ship and gave a hat and shirt and some wine and meat. According to Barlowe, the Indian liked these so much that shortly afterwards he reciprocated with a generous gift of fish which he had scooped up from a nearby creek. The next day the local Indian king, Wingana, came to welcome the English warmly.

The chief pilot of the expedition was Simon Fernandes, whose career symbolized the transfer of navigational expertise from the Iberian countries to their northern rivals. Fernandes was born in 1538 in Portugal, where he was trained as a pilot before going to work for the Spanish. While in the service of Spain he acquired complete information on its shipping and trade routes. He turned this knowledge to his own advantage by assisting English pirates to raid the Spanish ships and then, flagrantly avoiding customs, sold his treasure in England. The Spanish Ambassador to London reported to Philip II that Fernandes was "a thorough-paced scoundrel who has been giving them much information about that coast (the West Indies) which he knows very well. As I am told, he has done no little damage to the King of Portugal, by reason of the losses suffered in this kingdom by his subjects." Fernandes' activities became so notorious that he was tried in London in 1577 by the High Court of the Admiralty. But, despite one charge that he had murdered seven sailors during the seizure of a Portuguese caravel, he was freed. His skills were too important for the English to lose and in 1584 he entered the service of Sir Francis Walsingham, the Secretary of State to Queen Elizabeth and supporter of the Virginia voyages.

The following year, Fernandes guided the first group of English settlers to the island of Roanoke, which shelters inside the North Carolina Outer Banks. It was an inauspicious entry, for Fernandes grounded the expedition's flagship, the *Tiger*, while he tried to steer it through an inlet. The narrow inlets in the Outer Banks and the shallow waters which separate the islands within had confused even England's most skilled navigator. Roanoke turned out to be inhospitable, stagnant, marshy and disease-ridden. The next year most of the settlers returned to England with Sir Francis Drake, who had put in to Roanoke on the way back from the West Indies.

In 1587 the English tried once again under the helmsmanship of Fernandes. His orders were to pick up the holding force left on Roanoke the previous year and take them and the new settlers to a more hospitable site on the Chesapeake river. However, the holding force had left the island in a boat which was lost on the way back to England. Fernandes was so eager to sail down to the West Indies for some piracy that he refused to take his human cargo any further and dumped them on Roanoke. In despair the settlers sent their governor, John White, on a small boat back to England to bring relief. The war with Spain was at its peak and White could not return to Roanoke until 1590. He arrived to find that all the settlers had disappeared. They had probably tried to move north into the Chesapeake area but no remains were ever found.

Sir Walter Raleigh was the driving force behind the first English attempt to settle America. The Roanoke expedition sailed in his colors; and Raleigh laid plans for taking over the continent with a clique of ambitious young Elizabethans from the Middle Temple Inn of Court in London.

This tragic failure was relieved by one outstanding achievement, John White's illustrations and maps. White was an outstanding watercolorist and drew every facet of Indian life—villages, dress, dances and cuisine. He also painted the huge local variety of fish, birds and crustaceans. One of his illustrations showed a giant turtle, but little did White know, for it would have occurred to no Englishman to ask the question, that this was an important symbol in the creation myths of some Indians. One such myth told how a giant turtle emerged from the primeval ocean with a lump of mud on its back which transformed itself into a vast island on which the Indians lived.

What was the Indian name for this great island? The simple answer was "Ours." While the Europeans saw the land as something to be delineated, possessed and exploited, the indigenous peoples felt a more reciprocal relationship with the land, for it sustained them as they cared for it. William Boelhower in *Through a Glass Darkly* neatly makes the comparison: "Because he lived in nature, the Indian too would have to be subdued. The intrinsic desire of the mapper is to produce a perfect transcription of the land. . . . If the Indian protested, saying 'I am where my body is,' the colonist answered, 'I am where my boundaries are.'"

John White's map of part of Virginia, dating from 1585, marked the beginning of the dismissal by the English of the Indian view of their land. On one level it was the most accurate and detailed representation so far of any part of the American coast,

John White's map of Virginia. The painting is of outstanding quality, its accuracy largely determined by Thomas Hariot, the surveyor with whom White worked. Sir Walter Raleigh's arms are prominently displayed, a stamp of English possession imposed on the Indians.

delicately outlined by White's skilful hand. But its dominating symbol in deep red is Sir Walter Raleigh's coat of arms—a message to other entrepreneurs that here was a land for the taking. This meaning would have been all the more clear to the Elizabethans back in England because by now they were growing accustomed to view their own country and territorial claims within it as defined by maps. Christopher Saxton had carried out the first systematic survey of the counties of England and Wales between 1574 and 1579. It was made possible only by the patronage of Thomas Seckford and the support of Queen Elizabeth I herself and the Privy Council, which ordered local officials throughout the country "to see him [Saxton] conducted unto any towre, castle, high place or hill, to view that country . . . and that at his departure from any towne or place that he hath taken the view of, the said towne do set forth a horseman that can speak both Welsh and English to safe-conduct him to the market-towne." The completed survey contained thirty-five colored maps and was called *The Elizabethan Atlas*.

The Queen's first minister, Lord Burghley, was fully aware of the administrative importance of such a map in giving the government a full view of its domain. On his own collection of Saxton maps one annotation referring to "places of descente" reveals Burghley's anxieties about the constant threat of invasion from Europe. In addition to national boundaries, maps also defined individual estates and produced a more formal demarcation of ownership. In 1607 James I ordered the "Great Survey" of the Crown lands. Surveyors such as John Norden precisely outlined the great estates, including Windsor Park. One by-product of this new accuracy was that many tenant-farmers suddenly found themselves deprived of part of their lands. Such deprivation was nothing compared with that about to be inflicted on the Indians of North America. At the same time as James I ordered the survey of his lands, the second English attempt to settle in America was launched. Three ships, the *Discovery*, the *Susan Constant* and the *Godspeed*, arrived on the shores of the James river in Virginia and pitched camp. Some of those on board were craftsmen and

The loggerhead turtle, one of many American species drawn by John White at Roanoke. The turtle was an important American Indian symbol. In one Indian creation story, a being descended into the ocean, landed on a turtle's back and placed mud on it. This earth then turned into the great island of the world, supported by the turtle.

farmers, but many were gentlemen whose idea of settlement consisted of idling while new wealth accumulated around them. Such ambitions were soon dispelled by the unaccustomed hardship of living off the land and by the lack of gold and silver, the traditional easy road to riches.

The colony was sustained only by the ruthless leadership of John Smith, a mercenary and adventurer who had been enlisted to lend military expertise to the expedition. Smith had had a colorful past, having once been the slave of a Turkish princess from whom he escaped to earn a living by fighting pirates for the Russians. During the journey across the Atlantic Smith had been placed under arrest for alleged mutiny, although this was a cover for a general dislike of his arrogance. However, during the first harsh winter Smith's forceful personality saved the colony. He insisted that no one should receive food or shelter unless he was willing to work, and the gentlemen had to obey the yeoman's son.

Smith began to map the local country, taking a small boat through its complex network of tributaries and rivers, and in 1612 produced his map of Virginia. He fully acknowledged the presence of Indians and informed his readers that "as far as you see the little Crosses on rivers, mountaines, or other places, have been discovered; the rest was had by information of the Savages, and are set down according to their instructions." The great majority of place-names on the map are Indian. Indeed the early mapping of Virginia, New England and Canada could not have been achieved without the help of local Indians, who carried the knowledge in their head or carved it onto birch bark. Soon, however, not only their contribution but their very existence would be wiped off the map. Sixty years later a new survey of Virginia and Maryland by Augustine Herrman replaced the Indian names with English ones. Smith himself had contributed to this process, for his 1616 map of New England included no acknowledgement of the Indians who had provided the necessary information. In 1677 an updated map of New England by John Foster used only English place-names, reflecting the Indians' gradual dispossession.

The Jamestown colony, the first permanent English settlement in America, had developed amicable relations with the local Indian king, Powhatan, largely because of the marriage of his daughter Pocohontas to the settler John Rolfe. It was Rolfe who brought prosperity to the colony by exploiting tobacco as a cash crop. Established by this unexpected saviour, the colony began to expand and encroach further and further on Indian territory. The tensions became unbearable and in 1619 the Indians massacred two-thirds of the settlers. Reinforcements were sent out from England and uneasy coexistence gave way to invasion. The Algonquin Indians died or retreated inland.

In New England the same process was underway. Professor J. B. Harley has written that "just as in the nineteenth century we witness the 'Scramble for Africa' with native territories carved up by European powers on maps, so too we can talk about aspects of an English scramble for seventeenth-century New England." One story concerning John Smith's map of New England illustrates this idea. In 1623 the Council for New England acquired "new Letters Patent from King James" so that territory could be allotted to twenty colonial speculators. The method employed to divide the land was simply to cut Smith's map into pieces. The lots were drawn at Greenwich, in London, and Smith wrote that in the absence of one of the speculators, the Duke of Buckingham, "His Majesty was graciously pleased to draw the first lot in his grace's behalf, which contained the eighth number or share."

In the following centuries the dismemberment of America continued. William Boelhower sees in the familiar grid pattern of the United States an important symbol

William Penn makes his 1683 treaty with the Indians of Pennsylvania, in a painting by Edward Hicks. An English Quaker who was continually at odds with the authorities at home, Penn was granted the land by the Crown in 1681, in settlement of a £15,000 loan made by his father to Charles II.

of this process. The forests were the traditional habitat of the Indians, impenetrable and hostile to the Europeans. The settlers' response was to impose order, particularly that of towns, on this alien environment. Thus the sweeping avenues of the new capital Williamsburg grew out of the early sprawl of Jamestown, and William Penn's 1683 plan for Philadelphia with its rectangular squares and intersecting rectilinear streets became the model for American cities. Even the states themselves would be divided up by neat, straight lines as the settlers pushed west.

Throughout the seventeenth century the new world of America aroused intense curiosity in England and created a new industry producing the only means by which armchair travellers could view the latest discoveries. Maps became a passion of kings and commoners alike. The most accurate representations of this emerging world were to be found in Spanish and Portuguese charts but these were available only to a privileged few. Most people depended on printed editions of Ptolemy into which the new lands were incorporated. But accuracy was not necessarily the priority of the commercial mapmaker. The preparation of a woodblock or a copperplate for a printed map was costly, and the frequent need to respond to changes in the size and outlines of the world ate into profits. In the middle of the sixteenth century the center of mapmaking moved from southern Europe to the Netherlands, a shift that led to innovation. In 1538 the Flemish mapmaker Gerhardus Mercator made a world map in which for the first time the terms North America and South America were

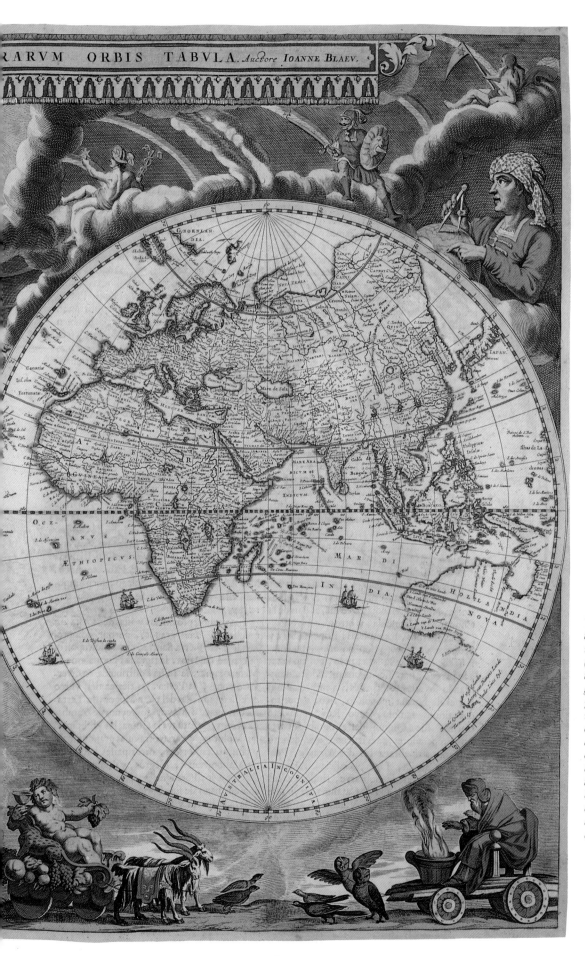

...RARVM ORBIS TABVLA. *Auctore* IOANNE BLAEV.

The two hemispheres of the world gloriously illustrated in a Joan (John) Blaeu map, one of the finest of the Dutch golden age. During this period, Amsterdam was not just the commercial capital of the world, but also produced the most ornate maps. Rival businesses were set up in the Kalverstraat and bitter battles fought in court as mapmakers accused each other of copying their plates.

A detail from a map of Sardinia in the Vatican map galleries. One gallery in the Pope's private quarters showed the two hemispheres of the world. The second, the Galleria delle Carte Geografiche was intended for public display and showed all the states of Italy. It was designed to impress upon visitors the Pope's authority in Italy at a time when it was under challenge.

included. But Mercator's fame would rest on his revolutionary projection. The problem of depicting a spherical world on a flat surface had assumed a greater urgency because of the use of sea charts over large areas of the world. In order to preserve the straight lines required to set compass courses the charts needed to compensate in some way for the curvature of the Earth. Mercator achieved this by increasing the distances between lines of latitude the further north or south you went. Mercator's projection became the standard format for world maps for over four hundred years, but although it served its navigational purpose, it distorted the proportions of the world. Lands in the far south or north such as Greenland appeared greatly elongated and those nearer the equator, such as India, appeared too small.

In 1570 Mercator's friend, Abraham Ortelius, published the first up-to-date atlas of the world since Ptolemy. Printed in Antwerp and called the *Theatrum Orbis Terrarum*, it comprised thirty-five leaves of text and fifty-three copperplate maps. It was an instant success and brought Ortelius a flood of praise. Mercator applauded in particular Ortelius' efforts "to bring out the geographical truth, which is so corrupted by mapmakers." Nevertheless accuracy remained elusive and some parts of the world were still subject to conjecture. Japan, for example, had a different shape on five different maps in the atlas and Ortelius preserved the great swathe of Ptolemy's *Terra Australis Incognita*.

Maps became an important part of culture and public display. The Vatican commissioned two beautiful map cycles. The world maps were painted in the Terza Loggia and symbolized God's rule over his creation. The maps in the Galleria delle Carte Geografiche showed all the states of Italy and were designed to impress visitors at a time when the Pope's secular authority faced constant challenge. These galleries were decorated less than one hundred years after Vasco da Gama's voyage to India and Columbus' discovery of America. In the sixteenth century geography and maps became a European obsession.

In a twist of history this European view of the world was now conveyed to the Chinese who had renounced the possibility of imposing their vision on everyone else. In 1583 the Italian Jesuit priest Matteo Ricci arrived in China to spread Christianity. Despite his strenuous efforts to learn Chinese and acquaint himself with local customs, Ricci fell under the suspicion of the Emperor. In the end, however, he managed to gain his favor by fascinating him with new European technologies such as mechanical clocks. Ricci informed the isolationist Chinese about the European discoveries and drew for them a succession of world maps. He tactfully maintained the goodwill of his hosts by placing China at the center. In the middle of the seventeenth century European knowledge arrived in Japan and the Japanese made some of the world's most beautiful maps on silk screens.

The pinnacle of European mapmaking was reached in the Netherlands in the middle of the seventeenth century. The Dutch Empire was itself founded on the back of Portuguese geographical advance. In 1593 two Dutch pilots, the brothers Cornelius and Frederic de Houtman, were despatched to Lisbon by the Calvinist minister Petrus Plancius to make covert enquiries about Portuguese navigation. They returned home with detailed charts of Portuguese possessions in the East Indies and a copy of the world map of the Pilot Major, Bartolomeo Lasso, which Plancius copied line for line. Further knowledge of Portuguese territory was brought home by the Dutch traveller, Jan Huygen van Linschoten, who had worked for the Bishop of Goa for five years, and now wrote his *Itinerario*, containing his detailed account of the Portuguese overseas. In 1597–98, de Houtman and his fleet made their first voyage around the Cape of Good Hope to the East Indies. The Dutch East

India Company was founded and started to expel the Portuguese from many of their most important overseas possessions and corner the spice trade.

Amsterdam became the mercantile headquarters of the Dutch Empire and, after the Spanish annexation of Antwerp, the new center of mapmaking. The Dutch East India Company attempted to keep secret the detailed charts of the Far East, but the production of world atlases became a highly competitive business. A leading proponent was Jodocius Hondius who had fled from Ghent in 1584 after it was stormed by the Spanish and lived in London for a while where he made a world map commemorating Drake's voyage around the world of 1577–80. He returned to Amsterdam and bought Mercator's copperplates from which he published an expanded edition of Mercator's *Atlas*. His son Hendrick and son-in-law Jan Jansson inherited the business but were eclipsed by the outstanding mapmaker of the age, Willem Blaeu. There was bitter rivalry between the two firms, each accusing the other of plagiarism and instigating a succession of lawsuits.

On Blaeu's death in 1637, his son Joan took over the family business and published his great *Atlas Major* in twelve volumes. By this time Dutch seamen had unintentionally provided a vital new piece of information about the world. Their route to the East Indies took them on a long stretch due east from the Cape of Good Hope before turning north towards Java. By this means they made the best of the Roaring Forties, the westerly winds in the southern hemisphere, rather than following the more tortuous Portuguese route by India. Because of the problems of calculating longitude, they sometimes missed the north turn and instead hit, on occasion with a lethal crash, the west coast of Australia. In 1642 the Dutchman Abel Tasman sailed round Australia and discovered the coast of New Zealand. It was a further piece of evidence that Ptolemy's great *Terra Australis Incognita* was an illusion. Joan Blaeu incorporated the information into his world maps. His most spectacular expression of mid-seventeenth-century Dutch confidence is his carving of the two hemispheres of the globe in the marble floor of Amsterdam's town hall. Today this imposing building is the Royal Palace, but three hundred years ago the burghers of Amsterdam could wander freely around this splendid hall and admire Blaeu's creation, the symbol of a world shaped, possessed, named and exploited by the nations of Europe.

6

TRIANGLES AND THEODOLITES

I n 1597, a century after Vasco da Gama's ships reached India, a less epoch-making fleet was approaching the English Channel from the Atlantic. The Earl of Cumberland, commanding, was returning home. An unnamed gentleman on board recorded the experience: "In the evening while yet it was fair his lordship commanded the lead should be wet, and at the second sounding, partly by the depth of the water and partly by the ground it was reasonably judged we were nearly entering the Sleeve [the Channel]. Marry, whether to the coast of France or our own coast there was difference of opinion."

To make matters worse, a storm blew up. "Our master would still have held the same course, north-east and by east, but his lordship about midnight absolutely commanded otherwise and gave instruction to sail a more northerly course, which the event showed was the saving of us all from utmost danger. For the next morning very early we saw land and quickly it was made Normandy, so that clear it is that when we began to alter our course we were exceeding near Ushant [Ile d'Ouessant] and the rocks, upon which if we had fallen in the night, there had been very little twixt us and sudden death."

Such incidents were commonplace in Elizabethan times—and indeed much later, right up to the late eighteenth century. Ships could often be as much as a hundred miles (160 km) from the position that dead reckoning—the "reasonable judgement" of this particular master—predicted. Accuracy was not much better on shore. Progress in fixing landmarks on the Earth's suface by their latitude and longitude, first suggested by Hipparchus and Ptolemy two thousand years before, had been slow; and that despite the fact that determining latitude was not difficult. The poet Chaucer knew how to do it, and even wrote a treatise on the astrolabe. You sighted the sun, either with an astrolabe, or a quadrant or other simple device such as a cross-staff, if possible at noon, and measured the angle of the sun's elevation above the horizon. After known mathematical adjustments, you looked up this measurement in tables of the sun's movements first prepared by the ancients, and found your latitude, generally expressed in degrees north or south of the equator. If the sun did not shine at noon, nor perhaps all day, you needed to set your sights on the stars, preferably the Pole Star. "You must tarry until night that some starre appeare,

Left A ship of the Spanish Armada founders off the coast of Ireland in 1588. Of the 130 ships sent against Britain by Spain, only 86 returned. Shipwreck at night, especially during storms, was an ever-present danger for navigators, who could not be sure of their exact position at sea until a reliable method of determining a ship's longitude was invented in the late eighteenth century.

A seaman's astrolabe, used to find a ship's latitude by measuring the angular height of the sun or a star above the horizon. The instrument is held aloft and the alidade or vane rotated until the sun or star is aligned in the sights. Its angle to the horizontal or vertical is then read off. Accurate measurement on a rolling deck at sea was almost impossible; an error of four or five degrees was normal.

which you perfectly know," advised the Elizabethan writer Thomas Blundeville. Again you referred to ancient almanacs to deduce your latitude.

Finding longitude was a different matter altogether. The sun was useless, and so were the stars. Nor could writers such as Blundeville offer help on land or at sea. The best measure of longitude remained Ptolemy's, and that was vitiated by inaccuracies. Ptolemy had used Pythagoras' theorem of right-angled triangles to calculate the difference in longitude between places A and B. The trouble was that the two "known" sides of his triangle were the distance along a parallel of latitude between the latitude of A and of B, and the distance between A and B as the crow flies (the hypotenuse), both of which were imprecise because of uncertainty about the shape and size of the Earth. Even assuming the Earth to be a sphere, as most thinkers now did, no one knew its exact circumference, or the length of a degree of latitude or longitude on the ground.

At the end of the Middle Ages mankind was scarcely further forward than Eratosthenes in his calculation of the circumference of the Earth more than fifteen hundred years before. But in the early sixteenth century the "figure of the Earth"—and thus the position of everything upon it—started to exercise men's minds. Some of the best of these, including that of Sir Isaac Newton, have refined that figure ever since. Geodesy, the science of the Earth's shape, has passed through three major periods in its history: from the ancient Greeks to the seventeenth century, when the Earth was taken to be a sphere; from the seventeenth to the nineteenth century, when it became a spheroid; and into the twentieth century, which regards the Earth as an infinitely complex figure, known as a geoid.

Jean Fernel has the honor of being the first to enter this field after the Greeks. His fame rests mainly on medicine—he was the greatest physiologist of the sixteenth century, "the modern Galen," physician to the king of France—but he was provoked by the exploration and discoveries taking place in foreign lands. "Plato, Aristotle and the old philosophers made progress," he wrote, "and Ptolemy added a great deal more. Yet, were one of them to return today, he would find geography changed beyond recognition. A new globe has been given to us by the navigators of our time."

Around 1525, when he was twenty-eight, Fernel decided to put the "old philosophers" to the test. He would measure a degree of latitude from Paris north to Amiens by travelling the distance in his coach. Its wheels would act as an odometer and he would then multiply a wheel's circumference by the number of its revolutions on the journey—17,024 as it turned out. The positions of the two cities, thought to lie about a degree apart on a meridian of longitude, he would determine exactly by observations with a quadrant. Fernel was fortunate: the twists and turns of his route, and its changes in elevation, seem to have been compensated for by the crudeness of his quadrant. The result was only a hundredth in excess of the true value of a degree of a meridian: ten or twenty times more accurate than Eratosthenes' value (and 250 times better than Columbus'). But Fernel had no way of checking his small experiment and it was forgotten, for the world was unprepared for such rigor.

The need to measure longitude was beginning to be felt though. The year before Fernel's expedition to Amiens, a commission had been summoned to attempt finally to establish the line of demarcation between the Spanish and Portuguese empires, an aim set out in Pope Alexander VI's post-Columbus Bull of 1493. The commission had broken up in confusion, for there was no means of specifying the line on opposite sides of the world, since it passed over ocean, in the Atlantic and in the Pacific. What did "100 leagues (300 miles/480 km) west of the Azores" convey, when a ship was unable to be sure of its position within 30 leagues (90 miles/140 km) in either direction? Even

The frontispiece of Galileo's Dialogue on the Two Chief World Systems, *published in 1632. As one of the architects of modern astronomy, Galileo's observations and theoretical works were of crucial importance for the measurement of the Earth and of its position in the universe; but his view that the Earth circled the sun was held heretical by the Roman Catholic church, and was stifled in much of Europe.*

The lens used by Galileo to discover the moons of Jupiter in 1610. Years of patient squinting through this lens yielded a set of tables of Jupiter's moons that could be used as a celestial clock in determining longitude. He broke the lens himself while handing it to a member of the Medici family, his patrons.

if a captain had been able to measure his longitude, he would have been unable to translate his result from degrees into leagues, since the length of a degree was uncertain.

Despair over longitude grew throughout the sixteenth century. Portugal and Venice offered handsome prizes for its "discovery," as did Philip III of Spain, whose father had lost an Armada in the seas around Britain. In 1598 Philip offered a perpetual pension of 6000 ducats plus a life pension of 2000 ducats and an additional gratuity of 1000 ducats. Wild schemes proliferated at court. Philip was soon so bored by the whole problem that when an Italian called Galileo Galilei sent his idea in 1616, the king was unmoved. A sporadic correspondence lasting sixteen years eventually put Galileo off too, and he took his idea to the Dutch instead. The States General and its four commissioners appointed to vet proposals, including the mapmaker Willem Blaeu, were impressed. They awarded Galileo a golden chain and one of them prepared to visit him, but the trip had to be cancelled when the Holy Office heard rumors of it. By the time negotiations were resumed, in 1641, it was too late; Galileo, already blind, died soon after.

But Galileo's method originated a technique of measuring longitude on land that became standard for the next two centuries. The secret lay in Jupiter's four moons, which Galileo had first observed in 1610 through his new telescope. Remembering his patron Cosimo de' Medici, Galileo named them "Cosmian Stars" (*Sidera Medicea*). As he squinted he could just see the moons appearing and disappearing on either side of Jupiter in a predictable pattern. With great care and infinite patience he drew up tables of their behavior over many years and it was these that he offered to potential purchasers of longitude measurement. The tables could be used as a celestial time-check, Galileo suggested. An observer wishing to know his longitude need only compare his local time, derived from the sun or stars, with the time given by the tables for Jupiter's moons, and he could calculate his position east or west of Galileo's observatory in Italy.

One may wonder why such a laborious method of timekeeping was necessary; were there no clocks of sufficient accuracy? There were not, and even the best watches available to Galileo lost or gained as much as fifteen minutes a day. The first timepiece to approach modern standards—and sound the death-knell of the sundial—was a pendulum clock made in 1657 by the Dutchman Christian Huygens, probably the greatest continental European scientist of the century. He had developed it from studies of the pendulum and its period of swing with different bobs, first undertaken by Galileo.

From the perspective of the late twentieth century, ever watchful of the time, taking for granted precision in all things, we find it hard to conceive the scientific world of the seventeenth century. It is familiar to us, and yet also strange. Medieval science, with its predilection for finding facts to fit ancient theories, was breaking up, although very slowly. Despite the heliocentric theory of Copernicus published in 1543, books assuming the Earth to be the center of the universe appeared until after the end of the *seventeenth* century. Astronomy and astrology were still intimately connected. Even Newton, one of the greatest scientists of all time, pursued alchemy in parallel with the study of gravity, gaining insights that led to his shattering work *Principia Mathematica*.

At the beginning of the seventeenth century the theories of Aristotle reigned supreme; a "vital principle" was thought to animate the whole of Nature. People believed that water could be condensed into earth or rarefied into air. Motion was as much the growth of a boy into a young man as the fall of a heavy object when

William Blake's Newton, *dated 1795, depicts the great scientist as a godlike figure, drawing on a chart beside his underwater grotto. "I don't know what I may seem to the world," Newton said near the end of his life, "but, as to myself, I seem to have been only like a boy playing on the sea shore, and diverting myself in now and then finding a smoother pebble or a prettier shell than ordinary, whilst the great ocean of truth lay all undiscovered before me."*

dropped. The reason why the object fell was that it had a "natural sympathy" for the center of the Earth. If it went upwards or sideways this was "violent motion": an agent had to propel the object by coming into contact with it; "action at a distance"—that is, gravity—was regarded as impossible.

Galileo, and Newton after him, rejected this vague concept of motion in his studies of projectiles, concluding that an object in motion was not altered in itself by its passage: Aristotle's "natural" and "violent" motion were the same. Gravity was the reason why objects returned to Earth, but Galileo did not pretend to understand its cause. He thus inaugurated a modern conception of science: that Nature may be studied and analysed without necessarily being explicable. As Newton later remarked when pressed to explain how a force such as gravity could be capable of acting without physical contact: "To us it is enough that gravity does really exist, and acts according to laws which we have explained, and abundantly serves to account for all the motions of the celestial bodies and of our sea."

Astronomy and mechanics excepted, science was devoid of mathematics until as late as 1800. The revolutionary instruments of the period, the first of which was the telescope, developed slowly as the idea of quantitative measurement assumed an importance it had never merited in the medieval mind. Balances, for example, which alchemists had long used for their arcane preparations, were initially used by scientists simply to establish whether a chemical reaction had led to a loss or gain in weight, not to measure such changes. Similarly, the barometer was used simply to demonstrate the existence of atmospheric pressure and to create small evacuated spaces for experiments. Twenty years went by before the correlation of height and pressure was discovered and calibration considered, to give a "history of the weather." The thermoscope, in which air in a bulb expanded when heated and forced up a column of water, was invented by Galileo in 1600; but the liquid thermometer calibrated with two fixed points did not arrive until 1665, the Fahrenheit and Celsius scales not until fifty and eighty years after that.

Even in mathematics progress was slow. We can date to the late Middle Ages the sine and cosine tables that would make trigonometrical surveying feasible, logarithms to only 1614. Around the latter date Arabic numerals became stabilized in their modern form, but Roman numerals were still commonly employed, especially in accounts. Modern symbols for addition, subtraction, division and so forth, did not become fixed until the second half of the seventeenth century, the time of Newton, when algebraic notation was also settled. Until then mathematical arguments were conducted in what the historian of science A. R. Hall has termed "diffuse rhetorical form."

Newton's controversial reflections on gravity in the 1670s had an immediate practical consequence for mapmakers, for he said that the Earth was not a sphere. Centrifugal force, caused by Earth spinning on its axis, was balanced by gravitational force, but this did not apply evenly to the surface of the globe. The equator moved faster than the polar regions. The equator must therefore bulge slightly, while the poles must be flattened, making an oblate spheroid. Gravitational attraction at the equator should also be slightly less than at the poles, since it weakens with distance from the center of the Earth.

This theory would soon be put on trial in France and remain there for sixty years while French and British scientists—"philosophers" in the language of the day—argued over the evidence. It had fallen on fertile territory, for the second half of the seventeenth century was the golden age of scientific societies, formed in reaction to the Aristotelian regime still prevailing in the universities. Experimentation and mathematics were exalted, tradition dethroned. In Britain it was the time of Robert Boyle, Robert Hooke and Newton, in Europe of Christian Huygens and Gottfried Wilhelm von Leibniz. In 1645 the nucleus of the Royal Society had come into existence in London as a meeting-place for philosophers wishing to discuss "new or experimental philosophy." Among those later to take part were Sir Christopher Wren and Samuel Pepys. In 1662 the society was incorporated under a royal charter, but left without government support. It was poor but free from political interference.

The equivalent in France was the precise reverse: wealthy but under the thumb of the king. The Académie Royale des Sciences (now the Institut de France) was founded in 1666 by Louis XIV at the instigation of his ambitious Minister of Home Affairs Jean-Baptiste Colbert, the power behind the throne. Colbert intended to attract the most brilliant scientific minds in Europe to carry out research, but never lost sight of his Majesty's avowed aim for the Académie: the improvement of the maps and sailing charts of the kingdom. This aim in turn required better astronomy. Colbert persuaded the king to give money for a grand Observatory to be sited at Faubourg St Jacques outside Paris, away from the distractions of the city. Ample laboratory space would exist for observations and research and for the residences of the astronomers. Claude Perrault, who had designed Louis XIV's palace at Versailles, would be the architect. The official meridian of Paris would run through the Observatory. In fact it remained the basis of French maps until 1913, when the

French accepted the Greenwich meridian in exchange for British acquiescence in the metric system.

By 1669 star-gazing at the Observatory was in full swing. Louis' largesse and Colbert's acumen had lured its brightest luminary from under the nose of the Pope— Giovanni Domenico Cassini, then forty-three. At the age of twenty-five Cassini had been elected to the first chair of astronomy, at the University of Bologna. Since then, apart from some surveying of rivers and boundaries for the Pope and for the Senate of Bologna, and some sensational attempts at blood transfusion between animals, Cassini had devoted himself to Jupiter's satellites. His telescope was better than Galileo's and so was his timing of the pattern, using the recently invented pendulum clock. Colbert had already offered him membership of the Académie in 1665. When in 1668 Cassini brought out some impressive tables of Jupiter's moons, known as *Ephemerides*, Colbert determined to lure him to France.

Technically, Cassini was on loan; but Colbert soon induced him to stay in Paris, ignoring remonstrances to Louis by both the Pope and the Senate of Bologna. In 1673 Cassini changed his name to Jean-Dominique Cassini, and was on his way to founding a cartographic dynasty. By the time of the Revolution, more than a century later, four generations of Cassinis had mapped France, putting into practice the idea of Colbert, who for some years had been frustrated in his ambition to make Louis XIV's France the most advanced country in Europe. He wanted her to have roads, bridges, canals and dikes, but found no accurate maps on which to base his grand design. He now desired his brainchild, the Académie, to put its mind to remedying this lack.

The Academicians chose the Abbé Picard, France's most noted astronomer, to make a start by measuring an arc of a meridian. Picard decided to use the method of triangulation. That is, he would measure a baseline between two points on the ground as exactly as he could and then make this one side of a triangle whose third point would be a distant but clearly visible feature. The other two sides he would calculate, not measure, by observing the angles they made to his known side and then applying trigonometry. This would reveal exactly how far from either end of the baseline the distant feature lay. Once that was known, the process could be repeated with other features and a chain of triangles eventually extended across the country towards the surveyor's destination. If the baseline were also measured there, it would provide a check on the accuracy of the triangulation.

Neither the principle nor the practice was entirely original. The mathematician Gemma Frisius first reported triangulation in 1533 in a book published in Antwerp. In 1615 another mathematician, Willebrood Snell, ran thirty-three interconnecting triangles across 80 miles (130 km) of frozen meadows in Holland. Twenty years later a London mathematician, Richard Norwood, invented the chain for measuring baselines. He had the patience to use one to lay out the entire 180 miles (290 km) from London north to York, allowing for bends in the road. Solar observations then permitted him to calculate a remarkably accurate length for a meridional degree: 69.2 miles (111.4 km).

What was new in Picard's triangulation was the combination of precise measurement on the ground with precise observation of the sky. His expedition was a sure index of how science was moving. His quadrant was made of iron and was mounted on a heavy stand to reduce vibrations caused by the operator. Instead of pinholes for sighting there were telescopes with cross-hairs. Picard was in fact the first person to use telescopes in surveying. A 10-ft (3-m) telescope was used for the star sightings to determine the vertical, or zenith, in latitude determinations. Other

Jean-Dominique Cassini, founder of the family that mapped France. Italian-born, Cassini was a brilliant astronomer, first for the Pope and the Senate of Bologna, later for Louis XIV, the Sun King. Over the century leading up to the French Revolution four generations of Cassinis surveyed France and produced La Carte de Cassini, *the first scientifically conducted national survey.*

telescopes—one even longer—enabled Jupiter's moons to be seen clearly. Two pendulum clocks "marked the seconds with greater accuracy than most clocks mark the half hours." Picard's route lay from Malvoisine near Paris to a clock tower at Sourdon near Amiens, thus following some of Fernel's route of a century and a half earlier. He laid out his baseline, 7 miles (11 km) long, on the road between Paris and Fontainebleau, using well-seasoned, varnished wooden rods end to end. Thirteen triangles were measured to Amiens.

Picard's computed length for one meridional degree, 68.65 miles (110.46 km), was strikingly close to today's value. As we now know, and Newton predicted, the length differs at the equator and at the poles; but in the 1670s neither Picard nor the

The Paris Observatory, constructed between 1667 and 1672. Its first director was Jean-Dominique Cassini, who had been lured away from the Pope by Louis XIV and his Minister of Home Affairs Jean-Baptiste Colbert. Together they made the Observatory Europe's foremost center of astronomy and mapmaking. It was here, on 1 May 1682, that the king was shown the first scientific survey of France and discovered to his horror that the coastline had "shrunk" by more than a hundred miles (160 km) in places.

Académie accepted Newton's new theory, word of which was beginning to filter through from the Royal Society. Following the great Descartes, who ridiculed "action at a distance," they were reluctant even to admit the concept of gravity, let alone that it might vary. However, they could scarcely ignore two pieces of hard evidence that emerged at the time. Observations of Jupiter and Saturn showed that both planets were slightly flattened at the poles and bulging at the equator. Secondly, one of the Académie's own expeditions in 1673 to Cayenne, a French colony on the north-east coast of South America, revealed that a pendulum clock that kept good time in Paris lost two and a half minutes a day near the equator. Richer, the leader, put this down to weaker gravity in Cayenne than in Paris, but the meticulous Cassini, *de facto* director of the Observatory and the man behind the expedition, was not convinced. "It is suspected that this resulted from some error in the observations," he commented reprovingly. Ten years later another party sent out to Guadeloupe and Martinique by Cassini with minute instructions came back with a similar result. Cassini remained doubtful, but Newton, when he read the report, felt justified in claiming in the third book of his *Principia* that the pendulum variation confirmed his theory.

This theoretical issue apart, on the surveying front there had been further important advances. Following Picard's lead a survey of the environs of Paris had been completed, linked to his arc of a meridian. It was the first example of a sequence later applied in all great national surveys: first a trigonometrical skeleton anchored by astronomical observations and baselines, then a topographical survey putting flesh on the skeleton—roads, rivers, bridges, estates and so on. Scattered surveys, in the wake of Louis XIV's invasion of Holland in 1672, had given the Académie more experience. Then in 1679–81 Picard and La Hire had been able to conduct triangulations around most of the coast of France. The outline map produced by La

Measurement of the Paris meridian in the late 1660s. The measurement was inaugurated on the summer solstice, 21 June 1667, when members of the French Académie assembled at Faubourg St Jacques near Paris to "locate" the planned Paris Observatory. This remains the official meridian of Paris and was the reference point for all French maps until 1913, when the Greenwich meridian was adopted. The wooden rods used to measure the baseline, and towers for triangulation on surrounding hillocks, are clearly visible.

Jesuits in Siam observing an eclipse of the sun in April 1688. Information sent to Paris by Jesuits as far afield as China and South America enabled Jean-Dominique Cassini, who was the first director of the Paris Observatory, to compile his planisphère terrestre, *a world map that became a model for many maps in the following century. Careful astronomical observations allowed the fixing of latitudes and longitudes of exotic places for the first time.*

Hire in 1682 was a revelation. It shifted the western coastline of existing maps about $1\frac{1}{2}°$ of longitude east in relation to the Paris meridian, the southern coastline about $\frac{1}{2}°$ of latitude to the north. Brest moved 110 miles (177 km), Marseilles 40 miles (64 km). On 1 May 1682 Louis XIV paid a visit to his Observatory and was shown the map. "Your journey has cost me a major portion of my realm!" he exclaimed on seeing his truncated kingdom for the first time.

It says much for the new prestige of science in the late seventeenth century, and no doubt for Colbert's powers of persuasion, that the survey went on. The next year the king announced that the arc of the meridian would be extended to the south of France, eight times further than existed already. Cassini would direct, since Picard had died. But soon Colbert too died, and since his patronage had been pivotal the project was now shelved until the turn of the century.

Cassini began to concentrate on star observation and the production of the most accurate map of the world yet attempted: his *planisphère terrestre*, eventually published in 1696, which became a model for many of the better world maps of the next century. It was drawn on the third floor of one of the towers of the Observatory in the form of a circle 24 ft (7.3 m) in diameter. Meridians ran out from the center at 10° intervals like spokes on a wheel, while parallels of latitude formed concentric circles at the same interval. In the middle was a pin with a cord attached to it carrying a small rider for rapid, convenient "spotting" of places. The world's landmasses were inevitably highly distorted, but the latitude and longitude of every place certainly were not. Information poured in to Cassini from astronomers all over Europe, who used his revised *Ephemerides* of Jupiter to locate the longitude of

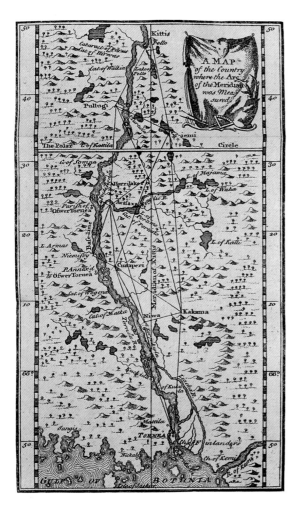

A map of Lapland compiled by the French scientist Maupertuis in 1736–37, to show his triangulation of a degree of a meridian. He began at Tornio (Torneå) at 65°50′N and stopped at Kittis Mountain just short of 66°50′N and just north of the Arctic Circle. His calculation, when compared with the length of a degree in France and on the equator in Peru, showed that the Earth was slightly flattened at the poles, as Newton's theory of gravitation had predicted.

hundreds of European towns and cities. The length of the Mediterranean was calculated properly for the first time. From farther-flung areas came the observations of Jesuits. Egypt, South America, the West Indies, Madagascar, Siam and China all began to be fixed. Even the Sun King must have been excited by such a panorama.

The unique *planisphère* helped Cassini to get royal approval in 1700 to revive the national survey, the beginning of scientific cartography. By then he was over seventy and needed assistance from his son Jacques. They began work on Picard's grand strategy: a backbone of triangles from Dunkirk to the Pyrenees, topographic filling-in of detail, and determination of the figure of the Earth. Jean-Dominique Cassini saw only the beginning, if that: like Galileo he was blind when he died in 1712 at the age of eighty-seven.

In the early decades of the new century the survey's progress was slower than it need have been, because Jacques Cassini had become embroiled in controversy. He had convinced himself by comparing his own work and that of his father with Picard's further north that a meridional degree became shorter as one moved north, not longer as Newton predicted. The Earth was indeed a spheroid not a sphere, but prolate not oblate: flattened at the equator not at the poles. His father, more cautious, had attributed the difference to errors of observation. Further measurements in 1718, after Jean-Dominique's death, appeared to confirm Jacques' view, and he now suggested that only decisive tests near the equator and the Arctic Circle could settle the matter. The figure of the Earth had become a *cause célèbre*, the most debated scientific issue of the day, with French and British national pride at stake. To confuse the issue further, the distinguished Huygens, before he died in 1695, had lent his weight to Newton.

In 1734 Louis XV, himself an amateur scientist, was finally compelled to take up the gauntlet thrown down by Jacques Cassini. More measurements in northern France that year had again demonstrated that a meridional degree apparently became shorter as one moved north. The Académie was instructed to form two expeditions from its members and make them ready to depart for Peru and Lapland. The expedition to Peru left France first, in 1735, but it was the Lapland party that returned first and broke the news that Newton (who had died in 1727) had been correct. Pierre Louis de Maupertuis, who led the Lapland expedition, was a notable scientist in various fields, especially biology, and later President of Frederick the Great's Academy of Sciences in Berlin.

Maupertuis was thirty-eight when he set out for Lapland with three other Academicians and a priest who also was a corresponding member of the Académie. They were joined by a Swedish astronomer at Stockholm, where the Swedish king welcomed them but told "frightful stories" of what lay ahead. Undeterred, they and their instruments arrived in early summer at Tornio (Torneå), a bleak village at the end of the Gulf of Bothnia, not far short of the Arctic Circle. The spire of the village church would form the beginning of their *magnum opus*, the marshes of the Tornio river their avenue north, following roughly the 24° meridian and the Swedish-Finnish border. The mountaintops, thickly forested, would provide sites for their stations of triangulation. First they had to be cleared enough for stations to be visible at about 30 miles (48 km). Signals had to be created by building great hollow cones of logs stripped of their bark so that their whiteness showed. As the lumberjacking began, so did the biting of the Tornio flies, the first of many privations suffered by Maupertuis, whose natural habitat was a Parisian salon.

Eventually, on 30 July, they crossed the Arctic Circle and found, about 25 miles (40 km) north of it, Kittis Mountain, which Maupertuis thought perfect for the termination of his triangles. There he took sightings of a star in the Dragon constellation, which he repeated later at Tornio to work out his latitude. Tornio and Kittis lay 57′ 27″ apart—just short of a degree. It was now late October, the river was freezing up and the daylight fast diminishing. The monstrous flies had gone and the cold had come in their place, as the Academicians prepared for their ultimate measurement. On 21 December, the winter solstice, when the sun at Tornio peers briefly over the horizon at noon, they set off across the ice on sledges. Using lengthy rods of fir to form the baseline, they labored for four or five hours a day in pale twilight, helped only by the whiteness of the snow and a steady shower of meteors. The cold was so extreme that only brandy stayed liquid, but when they tried to swallow it their tongues and lips froze to the cup.

Maupertuis did not leave Lapland until early summer 1737, after rechecking his entire network of triangles, taking new star sightings, and falling in love with a young girl, whom he brought back to France. On 13 November he stood before the Académie and gave an account of his ordeal. Then he announced his crucial value for the length of a degree: 69.04 miles (111.09 km) in the Arctic Circle, compared to Picard's 68.65 miles (110.46 km) in France; "Whence it is evident, That the Earth is considerably flattened towards the Poles."

The news reached the other expedition across the world in the Andes as they were nearing the end of their triangulation. Not surprisingly their leader, Charles de la Condamine, had difficulty persuading his team to continue. Conditions were far tougher even than in Lapland and the whole enterprise had been plagued by the suspicion of the Spanish rulers and the Peruvians, as well as by serious disagreement between La Condamine and his colleague Pierre Bouguer, nominally co-leader. La Condamine, an explorer in the Middle East as well as a scientist, unlike Bouguer seems never to have doubted his mission.

The expedition was the first outside group to enter the Spanish empire since the Spanish conquest of South America. The king of Spain had agreed only because he owed his throne to the French. Sailing via Martinique and Santo Domingo, the

Lapland, at the head of the Gulf of Bothnia. Maupertuis and his party of French Academicians landed here in the summer of 1736 and moved north to the Arctic Circle, measuring a degree of a meridian. They stayed nearly a year, enduring both the biting of monstrous flies that drove even the Laplanders and their reindeer to stay by the sea, and the bleak, freezing gloom of winter.

Academicians arrived in March 1736 at their base, Quito, in what was then colonial Peru. The local welcome was warm, even heroic, to begin with. But neither the Spanish nor the Peruvians believed the Frenchmen were mad scientists: both were convinced that they were after gold. Relations reached such a pass that La Condamine had to suspend surveying and journey eight months to Lima and back to see the Viceroy and obtain written permission to continue.

The plan was to triangulate about three degrees of meridian from just north of Quito on the equator to Cuenca in the south. They would be working in the very shadow of Mounts Cotopaxi and Chimborazo; and as the Himalayan surveyors would discover more than a century later, such heights test human endurance to the very limit. Members of the party—only some of whom had experience of military life—fainted, vomited and suffered "little hemorrhages" of the lungs, reported Bouguer to the Académie. As they crawled up paths through country "to which even the inhabitants were strangers," the mules stopped for breath every seven or eight steps. Then there was the Andean weather: "We have sometimes been obliged to purchase, for a month and a half's patience, a single quarter of an hour of fine weather; and in one of these stations we have been longer detained than we should have been toiling through a whole meridian in Europe."

Among these rocky peaks Bouguer made a discovery that has given his name to science. He noticed that his pendulum swung more slowly than expected, even after his altitude (and hence greater distance from the Earth's center) had been allowed for. The huge mass of the Andes had led him to expect a greater gravitational attraction and a faster rate of swing. It looked to Bouguer as if the Andes must be built of less dense rocks than the plains down below. His was the first inkling of the true complexity of the Earth's shape: not strictly a spheroid as had been thought, but a figure of unpredictable surface structure. Among today's geodesists such headaches are known as Bouguer anomalies.

Against the odds, the arc was completed in early 1743, by Bouguer at the north end and La Condamine at the south. Simultaneous star sightings enabled them to calculate the difference in latitude of the two termini while eliminating the error caused by the gradual rotation of the heavens; a refinement unavailable to Maupertuis with his single survey party. (A century later, in India, the technique would be brought to perfection by George Everest.) The length of a Peruvian degree was 68.32 miles (109.92 km), 0.33 miles (0.5 km) less than Picard's value and 0.72 miles (1.2 km) less than Maupertuis," which Bouguer and La Condamine now knew. There was no room for further doubt about the validity of Newton's insight of seventy years earlier.

After the disappointment of being second with their result, disaster now befell the Peruvian party. They had already lost their youngest member, the nephew of the Académie's treasurer, through fever. Now, in quick succession, the expedition's surgeon died in a riot in the bullring in Cuenca, the draughtsman died in a fall, the botanist lost his mind after a servant mislaid five years' worth of specimens, and the two Spanish officers sent by the king began a lawsuit over the omission of their names from the commemorative pyramids erected at either end of the baseline at Quito. Bouguer and La Condamine, who had narrowly escaped the same fate as the surgeon by hiding in a church, were no longer on speaking terms. They returned separately to France—Bouguer in 1744, La Condamine the following year.

La Condamine's irrepressible curiosity had taken him home down the Amazon by raft, the first competent scientist to make the trip. On the way he measured rivers, took soundings and noted exotic flora and fauna, including his most famous

A section of La Carte de France, *based on* La Carte de Cassini, *the first scientifically conducted national survey. The official meridian of Paris, which runs through the Observatory on the southern edge of the city, is marked with a fine grey line.*

discovery, latex from the rubber tree. He also experimented with poisoned arrows and their popular antidotes, salt and sugar, concluding that the poison worked and the antidotes did not. Altogether, he offered the adulatory gentlemen scientists of the Académie their first glimpses of a new world.

The two extraordinary expeditions to the Arctic and the equator were enough, said Voltaire, "to flatten both the poles and the Cassinis." Not quite. While the academicians were off in foreign parts, Jacques Cassini (II) and his son César François (III) had completed a network of four hundred triangles and eighteen baselines, covering France entirely. By 1745 they had a map ready, though without much topographical detail. They followed this with a topographical map of the Low Countries in the wake of the French victory at Fontenoy. These maps so impressed Louis XV that he gave the go-ahead to a topographical survey of all France. The Cassinis began adding rivers, canals, towns, châteaux, vineyards, windmills and watermills, even gallows. But in 1756, the year of Jacques Cassini's death, the king ran out of money.

Now came the Cassini family's finest hour. With Louis' blessing and the subscriptions of many at his court and of others, César François (III) transferred the survey from royal to private hands. Publishing rights in the maps would now accrue to his investors; *La Carte de Cassini* could continue. Although César François did not quite live to see its completion—that fell to his son Jean-Dominique (IV)—it was he who made Colbert's century-old vision a reality. In due course he became Comte de Thury. The great work, a *chef-d'oeuvre de géodesie*, was ready just in time for the Revolution: 182 sheets on the scale 1:86,400 covering 12 × 12 yards (11 × 11 m). Its military, political and economic value was so obvious that the new government agreed to support publication costs and pay for revisions as they became necessary. But Jean-Dominique's triumph was nearly short-lived. In 1794, during the Terror,

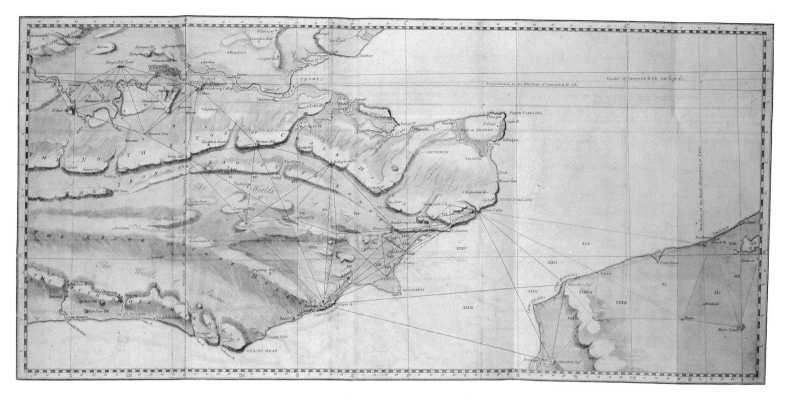

he was tried by a tribunal and spent six months in prison in uncertainty. His cousin, who had taken refuge in his house, was guillotined. Honored later by Napoleon Bonaparte, the fourth Cassini died in 1845 at the age of ninety-seven.

The year before the death of his father the Comte de Thury, the third Cassini had put forward a mildly revolutionary proposal of international significance. In 1783 the French ambassador to Britain conveyed it to Charles James Fox, one of George III's principal secretaries of state, who passed it to the Royal Society. Pointing out that the French and the British differed by nearly eleven seconds of longitude and fifteen seconds of latitude in their siting of the Paris and Greenwich Observatories, de Thury suggested a joint triangulation across the Sleeve (La Manche to the French; the English Channel to the British). Such a scheme would of course necessitate proper triangulation from Greenwich to Dover too.

The Royal Society was stung by the implication. But the facts were indisputable: in maps Britain was lagging behind France by many decades. And at least one of the Royal Society's members, General William Roy, was delighted that the French had now drawn attention to this. Roy had been acutely aware of the national failure since the 1740s, when he had roughly surveyed the Scottish Highlands after the Jacobite Rebellion. Ever afterwards he had been pressing the government to take action, but two major wars with the French and with the American colonies had delayed any serious move. In 1783, "for my private amusement," Roy had measured a baseline across the fields between Marylebone and St Pancras in north London and observed a series of triangles around the city. He had been about to deliver a paper on his hobby to the Royal Society when the third Cassini's memorandum had providentially turned up.

Roy readily accepted the challenge. From the outset he decided to use only the precisest instruments, constructed to his unique specification. A baseline was laid out on Hounslow Heath, southwest of London. The red Riga pine originally selected

A plan of the triangles measured by General William Roy between 1787 and 1790 to establish the precise distance between the Greenwich and Paris Observatories. Roy's baseline on Hounslow Heath, southwest of London, is marked, as are the official meridians of the two observatories. The two principal triangles across the English Channel, observed at night by limelight between stations at Dover Castle, Fairlight Head, Cap Blancnez and Montlambert, had sides as long as 45 miles (72 km). The survey marked the beginning of Britain's national survey, almost a century after France's.

for the rods proved too sensitive to changing humidity, and so glass tubes were substituted and sent to Jesse Ramsden, the leading instrument-maker of the day, to be made into 20-ft (6 m) lengths. Housed in wooden boxes, from which the ends protruded, the tubes were supported at each end and at three equally spaced points. One end of each tube carried a fixed stud, the other a spring-loaded stud with an ivory scale. When a line on the glass tube coincided with one on the ivory scale, the length was exactly 20 ft (6 m). Two thermometers with bulbs inside the box monitored temperature so that allowance could be made for expansion and contraction of the glass.

No baseline had ever received such loving care and minute attention. It was accurate to four decimal points. King George was among those who came from London to see it and he agreed to pay for the crafting of the theodolite to measure the triangles to come, on which General Roy had set his heart. It took Ramsden three years to create this device, which weighed 200 lb (91 kg) on completion and was accurate to within fractions of a second of arc: "a great theodolet," said Roy happily, "rendered extremely perfect."

At last, in autumn 1787, amid excitement, the two sides met at Dover, led by Jean-Dominique Cassini (IV) and General Roy. The meeting of minds was immediate and triangulation began across the water within days. Four stations were used—Dover Castle, Fairlight Head, Cap Blancnez, and Montlambert—and two principal triangles. The sides were long, as much as 45 miles (72 km), and it was necessary to take the observations at night using limelight. Fog and rain frustrated the surveyors but they persevered.

The scientific world had to wait nearly three years for the result, while Dover to Greenwich was scrupulously triangulated. Then General Roy could compute and present his findings. However, they were not notably accurate, partly because of continuing uncertainty over the exact figure of the Earth, and partly because of a lack of sophistication in Roy's mathematics. The whole campaign may have suffered somewhat as a result of Roy's becoming unhinged by Ramsden's long delay in delivering the theodolite; this may even have hastened Roy's death, which came suddenly while he was correcting the proofs of his report to the Royal Society of the "Channel crossing." The language he had used in it against Ramsden was "so violent that it had to be excised" by the Society. Nonetheless, "The first British triangulation had been carried out with vigour and skill," comments the *History of the Ordnance Survey*. Its real significance was symbolic. The next year, 1791, after Roy's death, George III established Britain's national survey on a proper footing.

France led the way in scientific cartography; Britain was next; and the Austrian and German states close behind. In the rest of Europe national surveys caught on during the first decades of the nineteenth century: Norway and Sweden in 1815, Russia in 1816, and Denmark and Switzerland in 1830. Beyond Europe, India excepted, the pace was slower, the task commensurately forbidding. A hundred years after the first country survey was completed, while the first faltering steps were being taken in Europe towards an International Map, nearly nine-tenths of the globe's land surface remained *terra incognita* to mapmakers.

The theodolite used by General William Roy to begin the national survey of Britain. Built by Jesse Ramsden, the leading instrument-maker of his day, the theodolite was capable of measuring both horizontal and vertical angles to within fractions of a second of arc. It weighed 200 lb (91 kg) and its graduated brass circle was 3 ft (nearly a metre) in diameter. To Roy it was a "great theodolet, rendered extremely perfect"; but he roundly cursed Ramsden for taking three years to create it.

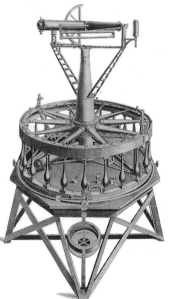

7

THE FREEDOM OF THE OCEANS

The land traveller in the second half of the eighteenth century could get a fair idea of where he was on the surface of the globe by observation of the sun and heavenly bodies. Not so the mariner. In spite of the courageous voyages and circumnavigations of the Age of Discovery, sea captains until the time of Captain Cook were often hundreds, sometimes many hundreds, of miles from where dead reckoning put them. There had been little progress in pinpointing position at sea in the three centuries since Columbus set sail for the New World.

The experience of a British naval officer, Commodore Anson, later First Lord of the British Admiralty, shows the risks sailors ran during long ocean voyages. In 1740 he sailed from Britain past Cape Horn to Manila and onwards back to Britain, capturing en route half a million pounds' worth of treasure from a Spanish galleon. Of the 1939 officers and men who embarked, 1051 perished, mostly from diseases such as scurvy caused by poor conditions aboard ship. Seventy or eighty of the victims died in the weeks after rounding Cape Horn, when Anson was unable to find the island of Juan Fernández, a well-known watering-hole for tired and sick crews.

To locate an island a navigator normally sailed along its latitude, which he found astronomically or from the sun's position, for as long as it took to sight his target. Seeking to save lives and time on this occasion, Anson sailed directly north for Juan Fernández as marked on his charts, but reached its given latitude without seeing land. He was then unsure whether to turn east or west. First he ran westward until (unknown to him) his ships lay within a few hours' sail of the island; then he turned east and ran all the way back to the coast of Chile. Finally, he turned and ran westward again.

The problem was that sun and stars could tell navigators their latitude north or south of the equator but not their easterly or westerly position. The search for longitude "overshadowed the life of every man afloat, and the safety of every ship and cargo," to quote a twentieth-century naval commander. Accurate measurement seemed an impossible dream, and "discovering the longitude" had become a stock phrase in the press, like "pigs might fly." Swift had used the idea in *Gulliver's Travels*; so had Hogarth, the foremost satirical artist of the age, whose *The Rake's Progress* shows a relatively sane-looking lunatic busy projecting longitude in a madhouse.

Cape Horn, the southernmost tip of South America. "The doubling of the Horn was a mighty thing indeed, and remained so until the end of the sailing ship era," wrote Alan Villiers in his biography of Captain Cook. Before the building of the Panama Canal, the route round Cape Horn (or through the treacherous Straits of Magellan) was the only western route for a ship from Europe to reach the Pacific.

A detail from The Rake's Progress *by Hogarth shows a lunatic drawing lines of longitude on the walls of a madhouse. The engraving was originally published in 1735, then republished in 1763, "retouched by the Author," just as a method for measuring longitude at sea was finally discovered. Obviously word had not yet reached Hogarth that the centuries-old problem had been solved.*

Taking precise measurements of Jupiter's satellites, which had solved the problem for the patient scientists of the Paris Observatory, was virtually impossible aboard ship. There was no way to keep the instruments steady, and besides, the satellites were often obscured by cloud when needed (and invisible in the daytime). A different solution was desperately called for, and the challenge attracted everyone from the greatest scientists—Galileo, Huygens, Hooke, Newton and Halley—to cranks. Large prizes were offered by the major nations of Europe.

The weirdest of the suggested solutions was an anonymous but apparently earnest proposal published in London in 1687. Its author began by citing a superstition that he admitted he used to believe: that a glass of water filled to the brim would run over at the instant of the full and the new moon. If this had been true, he remarked, it would have told the time accurately enough for a vessel's longitude to be found twice a month. Instead, he had come up with a better way, based on the then fashionable miracle cure of Sir Kenelm Digby, known as "powder of sympathy." Digby claimed to have stimulated a reaction in a patient at a considerable distance from himself, by immersing a bandage taken from the patient's wound into a basin of water containing some of the miraculous powder. Why not go further and send a wounded dog with every ship? At regular intervals suitably powdered bandages kept on shore could be dipped into water, causing the dog to yelp and act as an excellent timekeeper—much more convenient than Jupiter's moons!

More rational, but hardly practicable, was a method put forward in Britain in 1714 by Whiston and Ditton, men of some note in their day, the first a former Lucasian Professor of Mathematics at Cambridge (the post once held by Newton). They favored the establishment of permanent floating lightships on the principal trading routes. At intervals these would fire star-shells to a height of well over a mile (1.6 km). Ships could then determine their distance from the nearest lightship by timing the interval between the flash and the report. This would be fine if a captain could first find the lightship. Rather than a method for finding the longitude at sea, it was, Newton observed drily, useful "for keeping an Account of the Longitude . . . if at any time it should be lost."

The Académie in Paris had earlier rejected a device proposed to Louis XIV by an unnamed German inventor, which was attached to a ship itself. The king, without seeing the device, had already granted a patent and a cash payment of 60,000 livres (when Huygens was receiving 6000 a year!); subject to satisfactory demonstration of the invention, more money was promised. The device turned out to be a combination of waterwheel and odometer, to be inserted in a hole drilled in the keel of the ship. But the inventor could not really explain how this wheel would tell the difference between the wind-driven forward movement of a ship and the push and pull of ocean currents, in which the ship might be almost stationary relative to the water.

There seemed to be three serious contenders as solutions of the longitude problem: calculating the variation of the magnetic compass from true north, measuring lunar distances, and developing a marine chronometer for accurate timekeeping at sea. All three methods had influential advocates until the triumph of the chronometer in the 1770s.

Magnetic variation had first been noticed in the Age of Discovery. Until then, compass bearings and the charts based upon them referred unthinkingly to magnetic north rather than true north as measured by the elevation of the Pole Star. The distinction became crucial only when mariners began to use latitude as a way of fixing their courses. As they sailed along a latitude they observed that the compass needle unexpectedly varied in direction: it "northeasted" and "northwested." In other words, the direction of magnetic north shifted with the position of the ship.

The discovery initially caused dismay among the Portuguese pilots who first came across it. Having no idea of its true cause—the magnetic field of the Earth, which varies greatly in direction both from place to place and from decade to decade—they attributed the shift either to inferior lodestones, to badly touched or hung needles, or to leeway caused by ocean currents. Compass-makers began to make their own "allowances" for the variation by offsetting north on the compass card and providing different offsets for different parts of a voyage. Not surprisingly, Columbus found that his Flemish and Italian compasses did not read alike. There was much potential for perilous confusion.

But soon the idea arose that magnetic variation might be centered around a "true" meridian at which the variation from true north was zero. On either side of this meridian, it was thought, variation would increase uniformly in opposite directions—hence the "northeasting" and "northwesting." If that were so, the direction of magnetic north relative to true north could be predicted at any point around the Earth's circumference; and this would enable a ship to know its longitude by comparison with its measured north.

A mariner's compass, made in 1719. Accurate navigation was hampered by uncertainty about the variation of magnetic from true north. For a long time it was thought to vary regularly east and west of a 'true' magnetic meridian. We now know that it varies irregularly from place to place and year to year.

Unfortunately, it was not so. A famous Chief Pilot of the Portuguese Indian Fleet, John de Castro, was the first to establish that the variation seemed to follow no pattern, certainly not that of a "true" meridian. A century later, in 1638, a professor of mathematics in London named Henry Gellibrand confirmed that the variation was even less predictable than pilots had feared since it altered over time as well as place. This fact was nevertheless seized on by a teacher of navigation, Henry Bond. That year he predicted that the magnetic variation in London, then east of true north, would fall to zero by 1657 and then increase slowly to the west. When he was proved right, a royal commission of six, including Robert Hooke, examined the idea and approved it as a method of finding longitude. The commission's motives were more political than scientific, since Hooke at least knew the idea to be worthless; the commissioners were keen to spike the guns of a French pretender.

For whatever reason, the method refused to go away. In 1698 Edmund Halley, a friend of Newton and observer of the comet that bears his name, had himself commissioned captain of the *Paramour*, and sailed her throughout the Atlantic on the first-ever marine expedition for scientific purposes. Everywhere he measured the magnetic variation from true north, and then used his data to construct lines of equal variation, like contours on a relief map, for the year 1700. The map caused a stir in scientific circles and much scepticism among navigators, for Halley showed magnetic "north" lying east-west at the coast of what is now the United States, and a "true" meridian that looked more like a parabola than a straight line. Valuable though this research was, it ruled out magnetic variation as a practicable option for mariners needing their longitude, though it remained theoretically feasible given a wide enough spread of reference values such as Halley's, monitored from year to year (a condition not fulfilled until the use of satellites in the twentieth century). Halley did not abandon it, but he and other scientists increasingly put their faith in the method of lunar distances: measurements of the moon's position in relation to certain stars.

Unlike Jupiter's moons, Earth's moon changes position so clearly and substantially that if sailors could observe its movements accurately and compare them with lunar tables prepared in London or Paris, they could in theory calculate longitude. Much ingenuity was expended by inventors to help sailors achieve this desirable goal. But science in the seventeenth and early eighteenth centuries was simply not rigorous enough to overcome the obstacles to obtaining the necessary accuracy. In practice, as Newton wisely pointed out, the accuracy of lunar tables was between two and three degrees of longitude—no better than dead reckoning. Then, an angular error of $5'$ in reading a lunar distance meant an error of $2\frac{1}{2}°$ in the longitude, and such an error was only too easy to make. An imperfection in the observer's telescope, an error in the figures used to correct the observations for parallax or in the lunar tables issued by the Greenwich or Paris Observatories, or the inevitable rolling of the ship and the likely lack of training of the observer, would each be enough to make nonsense of the required accuracy of observation. Together they ensured that lunars were almost bound to fail, even if some considerable minds remained reluctant to admit it. Only in the late 1760s, with the publication of better lunar tables, did the method have a real chance of working; but by then it had been superseded.

That left the chronometer, which seemed scarcely more attainable a prospect. Seventeenth- and early-eighteenth-century watches might well lose or gain fifteen minutes a day, but a portable chronometer for measuring longitude, to be effective, could afford to lose or gain that much only over several years. As was recorded in the minutes of a British committee set up in 1714 to examine the whole vexing question of longitude, Isaac Newton stated that "for determining the Longitude at sea, there

Edmund Halley, the British Astronomer Royal from 1720 to 1742. Although he is more famous for his work as an astronomer, including his observation of the comet that bears his name, Halley was fascinated by the problem of longitude. His early advocacy of the marine chronometer made possible its eventual development by John Harrison.

have been several Projects, true in the Theory, but difficult to execute. One is, by a Watch to keep time exactly: But, by reason of the Motion of a Ship, the Variation of Heat and Cold, Wet and Dry, and the Difference of Gravity in different Latitudes, such a Watch hath not yet been made.''

The basic notion of using a chronometer to measure longitude dates from the thirteenth century. In 1522 it was stated clearly by Gemma Frisius, the mathematician who first conducted triangulation and whose work was translated into English and published in London in 1555. But at this time it was only theory, for the construction of such an accurate watch was well beyond the capabilities of the age. Not until the late 1650s was real progress achieved, by Huygens in Holland and Hooke in Britain, working independently. The challenge they both faced was to find a mechanical means to regulate motion as exactly as possible. While Huygens concentrated on using a pendulum, Hooke developed a revolutionary balance spring that coiled and uncoiled over an interval defined by the torsion of the spring and the size of the balance weights, obeying a principle we now know as Hooke's Law.

Both scientists' clocks worked well on land, though they proved much too temperature-sensitive for great accuracy, as a result of the contraction and expansion of the metal in the pendulum or spring. Only Huygens' clock was tested at sea. It did not prove satisfactory because the pendulum was affected by fluctuations of temperature, by the ship's motion, and by changes in gravity. (Observation of the latter fact helped to confirm Newton's law of universal gravitation.)

The man responsible for the next breathrough was of humble origin compared with Huygens and Hooke. John Harrison, born in 1693, was the son of a carpenter from Yorkshire. A severe attack of smallpox, when he was six, seems to have prompted his interest in watches, for as he lay recuperating he scrutinized the working of a watch

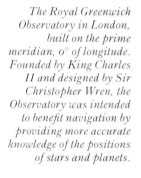

The Royal Greenwich Observatory in London, built on the prime meridian, o° of longitude. Founded by King Charles II and designed by Sir Christopher Wren, the Observatory was intended to benefit navigation by providing more accurate knowledge of the positions of stars and planets.

laid on his pillow. His education came from what he could pick up himself, including some mechanics and physics from a book lent him by a local clergyman. Harrison's first timepiece, which he built in 1715, was a grandfather clock similar in almost every respect to others of its day. But he soon turned to making more original and elaborate mechanisms. Examining a steeple clock that had stopped for lack of oil convinced him, it seems, that he must eliminate all friction from his timekeepers: his remedy, the "grasshopper escapement," was almost totally effective. Next he confronted the problems of temperature, and by making a pendulum out of different lengths of brass and steel, which expand differently when warmed and cooled, he created a pendulum whose length remained the same regardless of change in temperature, thereby disposing of that problem. By 1726 he had completed a clock that varied by no more than a second a month over the next fourteen years.

He was no doubt fully aware that in 1714, by Act of Queen Anne, a prize of £20,000 had been offered by Parliament to the inventor of a workable method of determining a ship's longitude to within half a degree (30'): that is, to within two minutes on the clock. A prize of £15,000 was offered for a determination to within 40' and one of £10,000 for a determination to within a degree. An august Board of Longitude, meeting three times a year at the Admiralty, was to administer the prize. Its members were to be the First Lord of the Admiralty, the Speaker of the House of Commons, the First Commissioner of the Navy, the first Commissioner of Trade, the Admirals of the Red, White and Blue Squadrons, the Master of Trinity House (the authority responsible for lighthouses and pilots' licences), the Astronomer Royal, and the Savilian, Lucasian and Plumian Professors of Mathematics. Nothing daunted, John Harrison journeyed to London in 1728 to lay before the Board samples and drawings of a marine chronometer. The Astronomer Royal, Edmund Halley, wisely advised Harrison not to rely on the Board for finance in building his chronometer, but to approach another Fellow of the Royal Society, "Honest" George Graham, Britain's leading horologist. Harrison called on Graham at ten o'clock one morning and left at eight after dinner, having received a most generous offer: Graham was prepared to advance the money, without interest or security, to build Harrison's No. 1 chronometer.

The project took six years in Barrow, where Harrison and his family now lived. Instead of a pendulum the chronometer had a balance spring with two 5-lb (2.3 kg) weights connected by brass arcs, so that their motions were always opposed. One would therefore counteract the other when a ship rolled, eliminating this source of inaccuracy. The chronometer, which weighed 72 lb (33 kg), beat seconds and showed seconds, minutes, hours and days. Harrison tried it on a barge and then took it to London, where he obtained a certificate signed by five Fellows of the Royal Society, led by Graham and Halley. The Board of Longitude agreed to a sea trial.

The First Lord of the Admiralty, Sir Charles Wager, wrote to Captain George Procter of the *Centurion* (the ship in which Commodore Anson would search for the island of Juan Fernández in 1741):

> Admiralty, 14th May 1736
>
> The Instrument which is put on Board your Ship, has been approved by all the Mathematicians in Town that have seen it (and few have not) to be the Best that has been made for measuring Time: how it will succeed at Sea, you will be partly the Judge: I have written to Sir John Norris, to desire him to send Home the Instrument, and the Maker of it (whom I think you have with you) by the first Ship that comes . . . The Man is said by those who know him

John Harrison, the Yorkshire clockmaker: "Inventor of the Compound Pendulum, and of Several Time-Keepers, for ascertaining the Longitude at Sea; The last of which, on a Voyage, ordered by the Commissioners of Longitude, was certified to have succeeded considerably within the Limits prescribed by Act of Parliament, of the 12th year of Queen Anne." The machine behind him (drawn much too large) is his No. 3 chronometer, that on his right hand the famous No. 4.

best to be a very ingenious and sober Man, and capable of finding out something more than he has already, if he can find Encouragement. I desire, therefore, that you will see the Man be used civilly, and that you will be as kind to him as you can.

Procter to Wager:

> *Centurion* at Spithead, 17th May 1736
>
> I am very much honoured with yours of the 14th, in Relation to the Instrument I carried out, and its Maker: the Instrument is placed in my Cabbin, for giving the Man all the Advantage that is possible for making his Observations, and I find him to be a very sober, a very industrious and withal a very modest Man, so that my good Wishes can't but attend him; but the Difficulty of measuring Time truly, when so many unequal Shocks, and Motions, stand in opposition to it, gives me concern for the honest Man, and makes me feel he has attempted Impossibilities; but Sir, I will do him all the Good, and give him all the Help, that is in my Power, and acquaint him with your Concern for his Success; and your care that he shall be well treated. . . .

Captain Procter died on arrival in Lisbon, so we do not have the benefit of his report on the No. 1 chronometer. But we do know that it behaved well on the way back and served to correct the landfall in Britain by nearly 100 miles (160 km). The journey was almost north-south, so there was little risk of the ship losing its longitude.

The Board now offered Harrison £500 to build No. 2, half of the payment to be made on delivery. When Harrison handed the new chronometer over in 1739 the Board decided not to test it, for fear it might be captured by Spanish warships. But they made available a further £500 towards the cost of No. 3. For reasons that remain unclear, Harrison took eighteen years, until 1757, to complete this instrument. In the meantime he was awarded the Copley Medal of the Royal Society, its highest honor.

On delivery of chronometer No. 3, Harrison immediately set to work on No. 4, a pocket-sized instrument, and made plans for No. 5, which was to be somewhat larger. No. 4, remarks the naval commander Rupert Gould, in his definitive study *The Marine Chronometer*, was a "very remarkable machine which, by reason alike of its beauty, its accuracy, and its historical interest, must take pride of place as the most famous chronometer that ever has been or ever will be made." It beat five to the second and went for 30 hours without winding. It weighed "less than the brain that conceived it." Externally, it looked like an exact reproduction, about 5 in (13 cm) in diameter, of the common pocket watch of the day; but its internal working was secret for over a century after its maker's death, in its details if not in its general principles. To Harrison, confiding in his journal, there was "neither any other Mechanical or Mathematical thing in the World that is more beautiful or curious in texture than this my watch or Time-keeper for the Longitude . . . and I heartily thank Almighty God that I have lived so long, as in some measure to complete it."

No. 4 left Britain in October 1761 on HMS *Deptford* bound for Jamaica and accompanied by Harrison's son William. On the ninth day William was able to advise the captain that they would sight the Madeira Islands the following morning. The captain offered five to one that he was wrong, but agreed to hold his course. The chronometer was correct. Two months later, at Jamaica, it was found to be just five

Harrison's No. 1 chronometer, completed in 1735. The bottom dial reads days, the right-hand dial hours, the left-hand dial minutes, and the top dial seconds. The instrument weighs 72 lb (33 kg).

seconds slow, corresponding to an error in longitude of a mere $1\frac{1}{4}'$, provided that the position of Jamaica had been correctly determined astronomically: well within the range specified to win the £20,000 prize.

But Harrison was forced to tussle with the Board for more than ten years to get his money. First there was a repeat sea trial to the West Indies, with excellent results. Even that would not satisfy powerful voices within the Board, some of whom, like the new Astronomer Royal Nevil Maskelyne, supported the method of lunar distances. More reasonably, the Board demanded to know how the chronometer worked, so that other watchmakers might duplicate Harrison's achievement. They were concerned that it should not be a "mechanical phoenix" that would die with its maker, writes Gould, "while Harrison was equally determined not to disclose its mechanism without payment to the last penny." Eventually, payment of a further instalment of £7500 being forthcoming from the Board, Harrison agreed to take the No. 4 apart in his house before a committee nominated by the Board, including the watchmaker Larcum Kendall. He also gave on oath a full explanation of the chronometer's mechanism and manufacture. But he did not yield an inch in his demand for the full £20,000.

By 1772 the row had become a scandal and George III decided he had to intervene personally. He had taken a liking to Harrison and had vetted his No. 5 watch—the Board still had charge of No. 4—in his private observatory at Kew. "By God, Harrison, I'll see you righted," the king promised. The royal intervention proved decisive. Three years before his death in 1776, "Longitude" Harrison received the full prize, by vote of Parliament. He also had the satisfaction of knowing that the No. 4 watch—or rather a copy of it made for the Board by Larcum Kendall—had received the supreme accolade from Captain James Cook, sailing in the South Seas on his second voyage, who in his journal referred to "our trusty guide, the Watch" that "exceeded the expectations of its most zealous advocate." Harrison's invention signalled the end of the pre-scientific era of navigation, though it took many decades before ships were equipped with reliable chronometers, partly because Harrison's mechanism was not suitable for mass reproduction and partly because of the conservatism of mariners. Nevertheless, "Landmark or no landmark," writes E. G. R. Taylor in her classic history of navigation, "the sailor knew precisely where he was—or had the means to know. He did indeed at long last possess the Haven-finding Art."

More than two centuries after the first navigation of the globe, huge tracts of ocean remained unexplored and even greater areas unmapped. The Pacific, one third of the Earth's surface, bigger than all the landmasses put together, was a vast mystery. Many scientists, following Aristotle and Ptolemy, still believed that an enormous Southern Continent, *Terra Australis Incognita*, awaited discovery in the high southern latitudes—a "balancing mass" for the continents of the northern hemisphere. Maps of the mid-eighteenth century often showed this continent stretching right across the bottom of the world. Although Magellan had never believed Tierra del Fuego to be a peninsula of this unknown land, many mapmakers, including the celebrated Mercator, ignored his view. Their faith was not shaken even by the discovery of western Australia by the Dutch in the early seventeenth century; it was not large enough nor far enough south to be the Southern Continent. Besides, it was too barren; believers were sure, as ever, that the new land would be fruitful.

According to one visionary, Alexander Dalrymple, ex–East India Company merchant, sometime seafarer to China, Fellow of the Royal Society, hydrographer,

King George III, painted by Allan Ramsay. The King championed John Harrison when the Commissioners withheld the £20,000 prize for developing an accurate marine chronometer; royal influence ensured that the money was eventually paid in full.

and author of two books "proving" the existence of *Terra Australis*, the Southern Continent had "a greater extent than the whole civilised part of Asia, from Turkey, to the eastern extremity of China. There is at present no trade from Europe thither, though the scraps from this table would be sufficient to maintain the power, dominion and sovereignty of Britain, by employing all its manufactures and ships." Dalrymple wanted to find the place of his dreams and become the Columbus of the Pacific. But the East India Company, preoccupied with India and the East Indies, was in no mood to expand, and the British Government strangely uncooperative.

In France, by contrast, the search for the Southern Continent had official backing. Bougainville had been dispatched to find it in the mid-1760s and had established the width of the Pacific by measuring a solar eclipse from one of the islands. (He had also unwittingly carried the first female circumnavigator: a French country girl who joined the ship as a servant, dressed as a man.) But like all the previous explorers of the Pacific—Magellan, Quiros, Torres, Tasman, Schouten, and le Maire, Roggeveen, Drake, Dampier, Byron, Carteret and Wallis—Bougainville had touched at a few islands, but left the ocean largely unexplored.

James Cook changed all that. In three daring voyages for the British Admiralty, from 1768 to 1779, he gave the world its first modern map of the Pacific, and became by general consent the greatest maritime explorer Britain has ever produced. His appointment was a bold one, for, like Harrison, Cook was a Yorkshireman without influence, the son of a farm bailiff. He had volunteered for the Navy after serving on the Whitby "cats": squat, steady boats ferrying coal up and down the east coast of Britain. While serving in Newfoundland he learned to survey and distinguished himself by some fine coastal surveys, including that of the St Lawrence estuary, becoming in the process "Mr. Cook, Engineer, Surveyor of Newfoundland and Labrador." His competence in all he undertook impressed his seniors.

Harrison's No. 4 chronometer, completed in 1761. A "very remarkable machine which, by reason alike of its beauty, its accuracy, and its historical interest, must take pride of place as the most famous chronometer that ever has been or ever will be made," according to Rupert Gould in The Marine Chronometer.

HMS *Endeavour*, a Whitby "cat" rigged as an ocean-going ship, sailed out of London on 21 July 1768, and headed out into the Atlantic. She was away for three years, during which time Cook lost not one man through sickness. In stark contrast to most eighteenth-century captains, Cook kept his ship as hygienic as possible, having it cured with fire at least twice a week. Not for Cook the silver-spoon test, in which foul air below decks was tolerated until a spoon tarnished badly when held up with the gunports and hatches closed. He stood no nonsense, "not even from bilge-water," writes the merchant sailing ship captain Alan Villiers, an Australian biographer of Cook. Food on the *Endeavour* was also better than the appalling rations on most vessels and so the dreaded scurvy was rare. A remarkably harmonious atmosphere prevailed despite the lack of protection against the polar cold and the congested living conditions so that Cook very seldom resorted to the typically brutal naval discipline of the time. The reasons seem to be his characteristic fairness and his knowledge of and respect for his ship.

On this first voyage such was the intransigence of the Board of Longitude that Cook did not have the benefit of Harrison's chronometer. Instead, when he reached Cape Horn, he struggled to take lunar distances, using improved tables issued the previous year by the Astronomer Royal Maskelyne. He soon appreciated how difficult lunars could be on a rolling deck. A lurch of the ship could easily send him flying, and midshipmen stood by to grab him and the instruments if necessary. The results were not bad: compared to the true values his latitude was out by only a mile (1.6 km), but his longitude erred by a degree, about 40 miles (64 km) at that latitude.

He was heading for Tahiti, where Bougainville had also called. Wallis, returning while *Endeavour* was being fitted out, had advised Cook of the island's charms,

Tahiti, which James Cook first sighted in 1769, after sailing along its known latitude for hundreds of miles until the island came into view. On later journeys, thanks to Harrison's chronometer, Cook was able to sail straight for Tahiti's known longitude.

causing him to change his plans. Rather than observe a transit of Venus, predicted for 3 June 1769, on the Marquesas Islands, he would view it from Tahiti. To find Wallis' island Cook was obliged to employ the same laborious method used for centuries: running down its latitude for hundred of miles until he sighted the island.

Tahiti was a revelation to Cook and his men—and to the Tahitians. Each had what the other most desired, but their encounter has rightly been called a fatal impact; for by their very presence the visitors helped to destroy a paradise. The same process recurred later throughout the South Seas, but in Tahiti more than anywhere it had the willing cooperation of the islanders. Their women exchanged sex freely and bountifully for goods, particularly iron. Soon both the Tahitians and Cook's own men were prizing nails out of the ship, threatening its strength. Cook was forced to issue rules, among them: "No sort of Iron, or anything that is made of Iron, or any sort of Cloth or other usefull or necessary articles, are to be given for any other thing but provisions.—J.C."

As far as we know, Cook himself remained chaste. But in his own undemonstrative way he responded strongly to the happiness of the Tahitians. No believer in the Noble Savage, despite the salon talk his voyages provoked in Europe, Cook developed a quick, sane concern for the future of the islands that today seems only too prescient. He wrote in his journal at the conclusion of his final call at Tahiti in 1777: It is to no purpose to tell them that you will not return. They think you must. ... I own I cannot avoid expressing as my real opinion that it would have been far better for these poor people never to have known our superiority in the accommodations and arts that make life comfortable, than after once knowing it, to be again left and abandoned in their original incapacity of improvement. Indeed they cannot be restored to that happy mediocrity in which they lived before we discovered them, if the intercourse between us should be discontinued. ... For, by the time that the iron tools, of which they are now possessed, are worn out, they will have almost lost the knowledge of their own. A stone hatchet is, at present, as rare a thing among them, as an iron one was eight years ago, and a chisel of bone or stone is not to be seen.

Similar doubts must often have pursued the practical, private Yorkshireman on his unique wanderings through this South Seas Serendip. But he was ever conscious too of the main purpose of his mission: to locate new lands in the south. Cook seems to have kept an open mind about the Southern Continent and about how close a ship might get to the South Pole. The officers and gentlemen on board the *Endeavour*, led by the naturalist Joseph Banks (later Sir Joseph and a long-time President of the Royal Society), were in friendly dispute: no-continent versus pro-continent. When land was sighted on 6 October Banks was sure it belonged to the Southern Continent. In fact it was the North Island of New Zealand, whose west coast Tasman had seen well over a century earlier.

Drawing on his survey experience in Newfoundland, Cook completed a 2400-mile (3860 km) circumnavigation in the difficult waters around New Zealand and made a magnificent map of its coastline, all in six months. "I doubt much whether the charts of our new French coasts are laid down with greater precision," remarked the French navigator Crozet at the time. Among the landmarks Cook named was a stormy headland on the north coast, where a rock was pierced through by the sea— Cape Brett, after Admiral Brett, whose Christian name was Piercey.

Cook landed more than once and his relations with the Maoris were good, as we might expect. "There was one supreme man in the ship," a Maori chief recalled fifty years later. "We knew that he was the chief of the whole by his perfect gentlemanly and noble demeanour. He seldom spoke, but some of his 'goblins' spoke much. He ... came to us and patted our cheeks and gently touched our heads.... My companions said: 'This is the leader, which is proved by his kindness to us; and he is also very fond of children. A noble man cannot be lost in a crowd.'"

The survey complete, Cook wanted to return to Europe by running before the wind east across the Pacific to Cape Horn between latitudes 40° and 60°, thereby establishing whether the legendary Southern Continent would really get in the way. Instead, bowing to a council of seamen, he sailed west and sighted the southeast corner of Australia on 20 April 1770. A curious unanimity that this was not *Terra Australis* prevailed aboard ship; even to Banks it looked too unproductive to fit his exotic image. *Endeavour* made her way slowly up the east coast of the subcontinent, the first European ship to do so. She stopped at what later became Botany Bay, where the Aborigines made the sailors unwelcome. Further up, in present-day Queensland, she was beached for much longer, while a gaping hole in her hull was repaired after the narrowest of escapes from the treacherous, pike-toothed coral of the Great Barrier Reef. Again, "All [the Aborigines] seemed to want was for us to be gone," noted Cook. Yet they seemed "far happier than we Europeans."

A year later the *Endeavour* reached Britain, where Cook's welcome was muted. Most of the acclaim went to Banks, a natural self-publicist with excellent

Captain James Cook by Nathaniel Dance, 1776. Cook had just returned from his second voyage to the Pacific, and was a celebrity. The portrait shows his public face, the image of Britain's greatest seafarer, who wrote that "the world will hardly admit of an excuse for a man leaving a coast unexplored he has once discovered."

connections. Dalrymple, unsurprisingly, was severely critical: yet again, he said, no one had looked properly for *Terra Australis*. As for Cook's "discoveries," they were already indicated in his own *Historical Collection of the Several Voyages and Discoveries in the South Pacific Ocean*. A second voyage was an absolute necessity, he added. On that at least, both Cook and the government agreed with Dalrymple.

HMS *Resolution* and *Adventure* sailed from Plymouth in July 1772 and this time on board *Resolution* was Kendall's copy of Harrison's No. 4 watch. It gave Cook the confidence to begin the accurate mapping of the Pacific. His principal aims on this second voyage were three: to investigate the Antarctic regions, to circumnavigate the globe at high southern latitudes, and to see as many new areas of the Pacific as possible. His instructions told him to "prosecute [his] discoveries as near to the South Pole as possible"; words upon which Cook was to place the bravest interpretation.

First he looked for the mysterious Cape Circumcision the Frenchman Bouvet claimed to have seen somewhere south of Africa. Perhaps it was a peninsula of the Southern Continent? According to the new chronometer, no land existed where Bouvet had marked it. Cook sailed on, into the Antarctic summer. Around his two fragile wooden ships, pitching and rolling, "the great 'bergs tossed and rolled all round, sometimes grinding together like the heads of fighting bull-elephants, cracking and banging with sounds like cannon-fire, with now and again one capsizing, rolling over in awesome, frightening grandeur, frothing the torn sea as if a meteor had suddenly entered it from space," writes Alan Villiers. The scene filled the mind "with admiration and horror," noted Cook.

Aboard, there was little insulation from numbing cold. Sailors could not wear gloves when adjusting the rigging aloft, because they would have interfered with the job. So their hands and fingers suffered fearfully from frost-burn and friction; nails were torn out at the roots, blood froze swiftly as it flowed. On Christmas Day 1772, as icicles dangled from every part of the rigging, the ship's company got drunk. Ice-fields appeared to block further advance but Cook persisted, at the risk of ice closing his escape route. On 17 January they crossed the Antarctic Circle and the next day— at latitude 67° 15', the farthest south reached by any man thus far—they turned round, faced by implacable pack-ice. Sailing north, away from the ice, out of the perpetual daylight, the sailors saw above them the strange, lurid display of the Southern Lights, so bright they could see their own shadows on deck.

The Southern Lights, or aurora australis, *over Adelaide Island in Antarctica. On Cook's second voyage the aurora was so bright overhead that his sailors could see their shadows on the deck of the ship.*

Continuing her circumnavigation of the polar regions, *Resolution* sailed on to New Zealand, where she linked up with *Adventure* again, having lost her in the Antarctic fogs. Mid-1773 saw Cook back in the tropics among the Pacific islands, before heading back to a new region of the Antarctic to take advantage of another summer. In January 1774 he came upon calmer, ice-free waters at longitude 110°. He sailed down and down in fog until 30 January, when he reached latitude 71° 10'. When the fog temporarily lifted, he saw pack-ice in his path and, on the southern horizon, great white clouds merging with icy mountains. Was this, finally, the Southern Continent? A frozen, uninhabitable wilderness? Cook had no way of telling for sure. (We now know—thanks to the accuracy with which Cook gave his position using the new chronometer—that he was still well over 100 miles (160 km) from the Antarctic coast.) Anyway, even he felt he had pushed far enough: "I will not say it was impossible anywhere to get further to the south, but the attempting it would have been a dangerous and rash enterprise, and what I believe no man in my situation would have thought of. It was indeed my opinion, as well as the opinion of

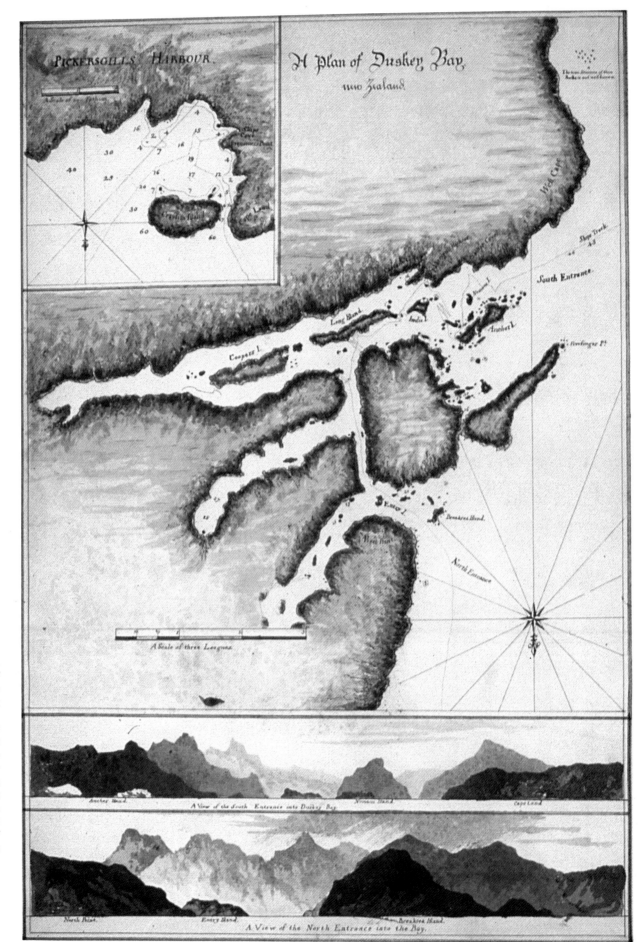

PICKERSGILLS HARBOUR.

A Plan of Duskey Bay, new Zealand.

A Scale of three Leagues.

A View of the South Entrance into Duskey Bay.

A View of the North Entrance into the Bay.

A detail from Captain James Cook's survey of New Zealand. In six months he completed a 2400-mile (3860-km) circumnavigation of New Zealand in difficult waters, and created a magnificent map of its coastline. The French navigator Crozet remarked that Cook's level of detail was as good as if not better than that of contemporary maps of the coast of France.

most on board, that this ice extended quite to the Pole or perhaps joins some land to which it had been fixed from the creation. ... I, who had ambition not only to go farther than anyone had done before, but as far as it was possible for man to go, was not sorry at meeting this interruption."

Sailing back to New Zealand in a great arc, taking in Easter Island and Tahiti, assisted as ever by the island-finding Harrison No. 4 chronometer, Cook now set course straight for Cape Horn before the west winds at latitude 55°, as he had failed to do on the first voyage. There was no trace of the bountiful *Terra Australis*. It was a myth. For the second time, Cook reached home three years after his departure.

Despite extinguishing the last hopes of a Southern Continent for Britain to trade with and colonize (and that just as the American colonies looked set to vanish), Cook's second voyage gave him the fame he deserved in London. The Royal Society awarded him its Copley Medal and he dined with the famous. But not everyone fell in with the general enthusiasm for things Polynesian. When Boswell told Dr Johnson that he felt "a strong inclination" to go with Captain Cook on his next voyage, Johnson replied, "Why, Sir, a man does feel so, till he considers how very little he can learn from such voyages." When the three volumes of Cook's *Voyages* were published, after his death, Johnson remarked: "There can be little entertainment in such books. One set of savages is very like another."

Dr Johnson, as is well known, could scarcely conceive of leaving London; Captain Cook found it hard to remain long in the city. Within weeks of his return he was planning a third voyage, his final one as it turned out. At the direction of the Admiralty, Cook was to search the northeast Pacific for a mirage as persistent as the

"The Santa Maria *beating through the Ice, with the* Discovery *in the most imminent danger in the distance." Cook arrived on the west coast of North America in 1778, on his third and final voyage, and proceeded slowly up to Alaska, charting as he went, eventually negotiating the Bering Strait and sailing almost as close to the North Pole as he had to the South Pole in 1774. Only solid pack-ice was enough to drive him back, on both occasions.*

South Continent: the Northwest Passage through the New World to China and the East Indies. Cook may not have had much faith that he would locate this, but he was happy only when he was exploring. In his view, "The world will hardly admit of an excuse for a man leaving a coast unexplored he has once discovered."

The British government still had sufficient faith in the existence of the Passage to have offered £20,000 in 1745 to anyone who found it. The theory was that it led somehow from the northwest corner of Hudson's Bay, via the Great Slave Lake and the Great Bear Lake, to the ice-free waters of the Pacific. No one knew that the Yukon stood in the way. British knowledge of the Pacific coast of North America stopped at Drake's New Albion, discovered two centuries before and never exactly located. In this way persisted the "great illusion, which for centuries held the minds of explorers spellbound," wrote the Arctic explorer Fridtjof Nansen in 1911. It would prove to be "no more than a vision, but a vision of greater worth than real knowledge; it lured discoverers farther and farther into the unknown world of ice; foot by foot, step by step, it was explored; man's comprehension of the Earth became extended and corrected; and the sea-power and imperial dominion of England drew its vigor from these dreams."

Cook was driven by similar motives. In 1778, *Resolution*, now accompanied by *Discovery*, sailed slowly up the west coast of Canada, charting as they went, and searching for an inlet that looked, from its tides, its fresh water, and other evidence, as if it might connect with Hudson's Bay 1500 miles (2410 km) to the east (a distance of which Cook was doubly certain, given the chronometer's accuracy). He was not hopeful, for at every stage massive mountains blocked any inland route. Oddly, he sailed past the west coast of Vancouver Island without realizing that it was an island. Once, Cook thought he had found the Passage and named it the "Gulf of Good Hope," but exploration by boat showed it to be the broad reach of waters leading to Anchorage and Mount McKinley in Alaska—today known as Cook Inlet.

At Unalaska, the peninsula that stretches out to the Aleutian Islands, Cook met a party of friendly Russian fur-traders. They had no knowledge of any Northwest Passage, nor any tradition of such an idea. But Cook pressed on through the Bering Strait as far as he dared among the grinding ice-floes, until he reached latitude 70° 44′—almost as close to the North Pole as he had been to the South four and a half years earlier. Before his ships could get trapped between the Siberian coast and the polar ice, he turned and retreated. It was just ten years since Cook had first set out; now his epic journeys were almost at an end. He died tragically in Hawaii, which he had discovered early in 1779, while attempting to calm an unnecessary fracas with the islanders. His chief memorial is the modern conception of the Pacific, derived from his superbly accurate charts, in which every island and every coastline had its latitude and longitude properly fixed for the first time through Cook's care and Harrison's chronometer. After Cook, no navigator could have an excuse for failing to find a Pacific island that Cook had visited or for being wrecked on a coastline appearing from nowhere. Next in importance came Cook's precious ecological record of the Pacific at a particular period, in specimens, drawings and descriptions, published in the best-selling *Voyages*. Lastly, there is an intangible memorial: Cook's enlightened spirit, his eighteenth-century view of human nature, epitomized by the names of his vessels—*Endeavour, Resolution, Adventure, Discovery*. "To these must be added a fifth," concludes Alan Villiers. "Its name is Humanity."

8

MEASURING INDIA

I ndia first took shape in a map drawn by Eratosthenes in the third century BC. He relied on observations made in Alexander's invasion of the subcontinent and on hearsay. The next attempt was by Ptolemy, in the second century AD. He had more to go on; the reports of Alexandrian merchants plying a flourishing trade across the Indian Ocean. Perhaps it was they who caused him to give pride of place to Ceylon, "Taprobane," and make it a huge island filling the southern portion of a moth-eaten Indian peninsula. But Ptolemy did get a remarkable amount right, including the existence of the Himalayas and the river Ganges flowing southeast from them towards the sea. His map remained the West's image of India until the sixteenth or seventeenth century.

Only then was a map drawn from measured routes and astronomical calculations. The work of a Jesuit, Father Monserrate, a guest of the Great Mughal Emperor Akbar, it went unpublished until this century. Another visitor at the Mughal court, the English ambassador Sir Thomas Roe, became the first European to give the contemporary West its first tolerably accurate picture of fabled Hindustan. Sailing back to England with the explorer William Baffin, Roe described his adventures to him, and Baffin reduced them to a map and published it in 1619. Beneath it he wrote the Latin legend *Vera quae visa; quae non, veriora*: "The things that we have seen are true; those that we have not seen are truer still."

The tag encapsulates the hopes of the ambitious merchants of the British East India Company, whose spearhead Roe was. A century and a half after Roe's embassy, in the 1750s, they and their French rivals were still largely ignorant of India's vast interior, 1800 miles (2900 km) in length and breadth. As late as 1752, the best available map of India—by the Frenchman d'Anville—is candidly blank within, showing just the coasts and rivers studded with names. The fact was that the European traders knew only their settlements on the coast; in the British case Calcutta, Madras and Bombay. From these, and from a few rivers, they did business at arm's length with the subcontinent, as branches of a Company with its headquarters in far-off London or Paris, not as imperial rulers.

Within a decade this utterly changed. In the east, near Calcutta, the East India Company's army under Robert Clive defeated the representative of the Mogul power

Robert Clive, Governor of Bengal, from a painting by Nathaniel Dance. Having laid the foundations of British rule in India, Clive saw the need for detailed surveys of the East India Company's territories, and in 1765 he ordered James Rennell to begin a survey of the Bengal Presidency.

Left *Fort St George, Madras, on the south-eastern coast of India, in the late eighteenth century. The fort dates back to 1639, when Britain's East India Company established a foothold on the subcontinent as it did at Bombay in the west and Calcutta in the northeast. Not far from the fort, in 1802, William Lambton laid out his first baseline for triangulation, and launched the Great Trigonometrical Survey of India, which his successors completed in the Himalayas some seventy years later.*

and acquired the Bengal Presidency, an enormous tract of land; in the south, a British victory drove the French out and secured the Madras Presidency. Very soon the Company demanded surveys of the new territories: first, to estimate their potential revenue, secondly to protect communications with the coast, and finally to defend them against attack.

One of the first to enter the field was an Englishman of twenty-two who had just left the navy in Madras, seeing no hope of promotion after the Peace of Paris had brought an end to the war with France. James Rennell's first assignment was typical of Company priorities in India. Take a budgerow, one of the lumbering keelless barges, he was told, and see if you can find a quicker way of shipping goods down the Ganges to Calcutta, since the existing route is unnavigable seven months of the year during the dry season. While you are at it, make a map of the Ganges delta.

Rennell's survey began unpromisingly: his budgerow sprang a leak and almost sank on the first night. But he got himself afloat and spent the next year taking soundings of river depth, getting compass bearings of features in the surrounding jungle and swamp, and making sketches. "We have no other Obstacles to carrying on our Business properly," he noted in his journal, "than the extensive thickets with which the Country abounds, and the constant dread of Tygers, whose Vicinity to us their Tracks, which we are constantly trampling over, do fully demonstrate." Rennell was then mapping the Sunderbans on the Bay of Bengal, haunt of the man-eating Royal Bengal Tiger. Not long after, a tiger carried off one of his Indian soldiers.

Nothing worse than "a Feaver" befell Rennell himself; though in Bengal, or indeed anywhere in India, a fever often led to death—a word spelled without a capital, so commonplace was it among Indian surveyors. Death by disease stalks the two thousand or so pages of the five volumes of the *Historical Records of the Survey of India* up to 1861, a monument to Himalayan human willpower, compiled by a later surveyor, R. H. Phillimore. It is no exaggeration to say that the slaughter on Survey was worse than in many famous battles of the time. (The mortality rate of ordinary Company men was bad enough. Scarcely one in seventy returned to England to enjoy his fortune, said Rennell.)

In mid-1765, while resting at Dacca, the main port of East Bengal (now Dhaka, capital of Bangladesh), Rennell received orders to continue his survey and map the entire Presidency of Bengal. They came from Lord Clive, the Governor General. Precision was not called for; Clive wanted to find out what the province was worth to the Company. Some latitude determinations were to be made, but not so that the overall map was delayed; longitude was hardly to be considered.

So Rennell set out on the first of a long series of special "route surveys": a term usually reserved in India for surveys attached to a military or political mission. This time he nearly died. Coming across an officer friend up-country, whose detachment was fighting a marauding group known as Sannyasis, Rennell received wounds that would impress even Clive: "I was now in a most shocking condition indeed, being deprived of the use of both my arms; a cut of sabre had cut through my right shoulder bone, and laid me open for nearly a foot down my back, cutting through and wounding some of my ribs." His head was also hurt. It took six days by boat to get him back to Dacca without medical assistance of any kind. Four years later Rennell thrust a bayonet down the throat of a leopard that had jumped him after maiming five of his men. No wonder Rennell lived to the ripe old age of 88.

Small wonder too that he decided to return to England as soon as he could complete his work and secure his future there. Clive had appointed him Surveyor

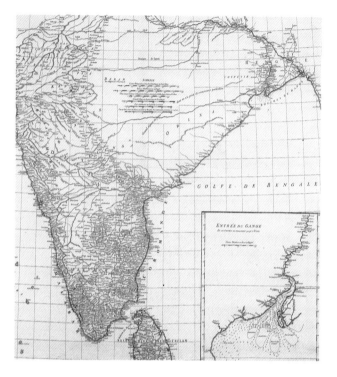

A French map of India, drawn by d'Anville in 1752. In the mid-eighteenth century the European colonists of India derived their knowledge of the country largely from their trading contacts along the coasts and rivers; no reliable information about the interior of the subcontinent was available. Hence d'Anville's candid comment written across central India: Grand espace de pays dont on n'a point de connaissance particulière—*"A great expanse of country about which we have no exact knowledge."*

General in Bengal in 1767; ten years later his *Bengal Atlas* was complete, compiled from five hundred surveys by himself and nine assistants. It was published in London, where Rennell was then living. He was already at work on a *Map of Hindoostan*, drawing upon the flood of information since d'Anville's honest effort of 1752. He had still to leave many blanks and much of the available data was discordant, but he had a gift for patient selection. The Himalayas on Rennell's map are better positioned than before; so is the lower Ganges, even if he persisted in believing Father Monserrate's notion that its source lay behind the Himalayas. (Ptolemy knew better.) Rennell himself had seen neither the source nor the mountains, but he had deduced their tremendous height after catching sight of peaks rising snow-capped from the plains of Bengal 150 miles (240 km) from his camp. He was the first European to be fascinated by the mysteries of the Himalayas.

Rennell's map was outdated almost as soon as it appeared—to his considerable pleasure. A second edition was published, and then a third in 1788, enlarged, along with his remarkable *Memoirs*. The latter went through two editions in the 1790s and was still in East India Company use in 1824, long after the map had been superseded. Throughout that period Rennell was known as the Father of Indian Geography, consulted by the Company on all matters geographical.

Each edition of the *Map of Hindoostan* showed more of India under Company control, and a greater increase in detail. The Honourable Court of Directors in London and their Council in India were no longer satisfied with rough-and-ready surveys by soldiers on the march. There was a pressing need for accuracy, which required the fixing of reference points in India by latitude and longitude. In 1792 the Company established its first observatory in India, at Madras. It also followed up a suggestion made in 1787 by General Roy, the founder of the Ordnance Survey in Britain, that chains be used to measure a meridional degree at lower latitudes. Although a baseline was laid out north of Calcutta the plan came to nothing. Part of the reason was shortage of instruments, which all had to come from Britain and took up to a year to reach India, and part the temperament of the man chosen for the job.

Reuben Burrow was an uncouth genius, one of the most intelligent and opinionated of the many men who mapped India. After schooling in Yorkshire, "interrupted by farm duties," followed by periods of assisting the Astronomer Royal, Nevil Maskelyne, in teaching mathematics to artillery cadets at the Tower of London, and editing the *Woman's Almanack* ("Aenigmas, Rebuses, and Mathematical Dissertations"), Burrow got fed up and sailed for Calcutta. Scathing from the outset about the imprecision of Rennell's Bengal surveys, he put forward a scheme for an astronomical survey of the Presidency, which was accepted by the Surveyor General.

Like Rennell, Burrow went by boat. Unlike Rennell, he complained loquaciously: "As I was taking equal latitudes a number of people came up and disturbed me by beating the ground and shaking the quicksilver; and soon after one of their Chief Officers came, apparently displeased, and inquisitive what I was about; I told him first to drive the people away and when I had done I would tell him; after concluding my observations I explained to him that I was correcting my watch; but he did not seem satisfied at all with my explanation; and I afterwards found that the Captain had told him inadvertently that I was a Conjurer." At Dacca "people were dying in heaps, and we were immediately almost taken sick; . . . and to add to our misfortunes, we had forgotten to wind the watches up." The twin devils of sickness and unreliable chronometers dogged Burrow's progress until his lonely death in his budgerow on the Ganges.

James Rennell, Surveyor General of Bengal. While mapping the province, Rennell came within an inch of death more than once, sustaining terrible sabre wounds that impressed even Clive. But he survived and produced his Bengal Atlas *in 1777, followed by three editions of his* Map of Hindoostan, *the first accurate map of India. Both were published in London, where Rennell settled after leaving Bengal.*

Burrow's latitudes and longitudes nevertheless remained the best available for some years. It was a pity that he died before he could carry out the Company's directions and become the first all-India surveyor by visiting Calcutta, Madras and Bombay and measuring their actual and relative positions—a sea journey of up to 2500 miles (4020 km). Probably, as the Directors feared, Burrow's vital timekeepers would have given trouble; but they would still have defined the growing empire more accurately than Rennell's *Map of Hindoostan*.

The great Indian surveys of the nineteenth century are the story of the indefatigable in pursuit of the unmeasurable, set in a land of notorious imprecision. The men that led them turned the pursuit into a marvellous obsession, egging on a Company that instinctively was "all for economy" and making do without detailed maps—just as Indian rulers had done for centuries before. But in 1800 the Company for once needed no encouragement. The previous year at Seringapatam in the south, its armies under Arthur Wellesley, future Duke of Wellington, had finally destroyed the dashing Tipu Sultan. Mysore, with an area almost the size of England, had come under Company control and required surveying, as did most of south India.

Two officers who had distinguished themselves against Tipu now stepped forward independently with plans. The first was Captain Colin Mackenzie, a friend of Wellesley, who proposed a detailed topographical survey going beyond mere military or geographical information to embrace a statistical account of the entire state of Mysore. He would be the first to employ accurate trigonometrical techniques backed up by astronomical observations; the hilly country of Mysore was well suited

A view of the fort at Ramnagar near Benares on the river Ganges, by Thomas and William Daniell, 1788. The boat on the left is a pinnace budgerow, a lumbering keelless barge of a kind much used by Europeans, surveyors included. James Rennell's water-borne survey of the lower reaches of the river in Bengal went by budgerow; Reuben Burrow, uncouth genius, died alone in one.

The siege at Seringapatam near Mysore in south India, 1799, where East India Company troops under Colonel Arthur Wellesley, later Duke of Wellington, destroyed Tipu Sultan of Mysore and annexed his territory. Two officers who distinguished themselves in the siege, Mackenzie and Lambton, now began the survey of south India.

to providing stations visible to each other, unlike the lush plains of Bengal. In early 1800 Mackenzie set out, assisted by two surveyors, a Dr Heyne in charge of botany, mineralogy and natural history, several boys from the Madras Observatory surveying school, and lascars and artificers to handle and maintain the equipment. For the next eight years Mackenzie was in the field.

Captain William Lambton, the second of the officer pioneers, was less well connected than Mackenzie and more introverted. "Some peculiarity of manner adhered to him from having lived so long out of the world," a fond assistant recalled after his death. It dated from the fifteen years Lambton had spent without promotion as Barrack Master in Nova Scotia, time in which Lambton had read the latest books on surveying and the infant science of geodesy. The works of the late General Roy and the Ordnance Survey were an inspiration to him. When he finally reached India he was in his mid-forties, an age when many old India hands were thinking of retirement.

Lambton's unique mastery of his subject explains why Wellesley and the hard-headed Directors gave support to his expensive, unlimited proposal. Lambton had in mind nothing less than the measurement of India belonging to the Company to the highest degree of precision possible. Beginning in the south, he intended to cover the subcontinent with triangles: a trigonometrical survey linked to a fixed point that would form a skeleton for all topographical surveys to come. More than that, he would "accomplish a desideratum still more sublime, viz. to determine by actual measurement the magnitude and figure of the Earth, an object of the utmost importance in the higher branches of mechanics and physical astronomy." Lambton had been fired by the experimental spirit of the age to amplify the discoveries of the French expeditions to Lapland and Peru sixty years before him: "Should the Earth prove to be neither an ellipsoid, nor a figure generated by any particular curve of known properties, but a figure whose meridional section is bounded by no law of curvature, then we can obtain nothing until we have an actual measurement."

Not everyone could be expected to appreciate the obscure captain's vision. James Rennell, for one, misunderstood and recommended the Company not to go ahead. The Company had informed him that the new survey was an astronomical survey and he imagined that Mackenzie's topographical survey, which he thoroughly approved of, was afterwards to be repeated in part by Lambton, with a completely independent set of astronomical observations. Rennell later graciously acknowledged his error when Maskelyne, the Astronomer Royal, had explained the idea properly. The members of the Finance Committee in Madras posed a more immediate difficulty: they plagued Lambton with questions and comments. One remarked: "If any traveller wishes to proceed to Seringapatam, he need only say so to his head palanquin bearer, and he will find his way to that place without having recourse to Captain Lambton's map." The captain gave as good as he got in these exchanges and eventually the Governor of Madras had to call a truce.

Lambton's first technical hurdle was to get hold of good instruments. He had a zenith sector for finding the angle of stars to the zenith, but lacked a chain and theodolite. A 100-foot (30-m) steel chain like General Roy's, with forty links of $2\frac{1}{2}$ ft (0.75 m) each, was located in Calcutta and purchased by the government; it had been rejected as a gift by the Emperor of China. Finding a suitable theodolite proved more difficult. A half-ton (508-kg) instrument, later known as the Great Theodolite, had to be specially constructed in London, and did not reach Lambton until 1802. A complimentary letter with it from the French Governor of Mauritius revealed that a French frigate had captured the instrument en route.

The early morning of 10 April 1802 found Lambton on a flat plain near Madras—like General Roy on Hounslow Heath fifteen years before—adjusting his chain in five wooden coffers supported on tripods with elevating screws and a weight at one end to keep the chain tight. It took him six weeks to measure a baseline $7\frac{1}{2}$ miles (12 km) long. Its longitude was determined by triangulation from the Madras Observatory, which thereby became the lynchpin of the entire Survey of India. The latitude was given by Lambton's careful observations with his zenith sector: sixteen nights devoted just to the star Aldebaran, for example. The Survey was ready to take to the field.

Lambton first drove the triangles across the peninsula from coast to coast. This was a shrewd calculation, quickly justifying the Survey in the eyes of its paymasters; by 1806 Lambton could already prove that the peninsula was about 360 miles (580 km) long on parallel 13°, not the 400 miles (640 km) deduced by Rennell. Unlike Louis XIV, who had been shocked by the shrinkage of his kingdom, the East India Company was pleased. Further north in Mysore, Mackenzie was less pleased. The parallel passed through the area of his survey, and he had hoped to beat Lambton to it despite his less accurate methods. But the rivalry that existed between these two differently talented men remained friendly, at least until near the end of Mackenzie's life.

Lambton now geared himself to net the whole southern portion of the peninsula with triangles: vertebrae on the spine that the Survey would extend along the 78° meridian from Cape Comorin to the foothills of the Himalayas. This spine would become a Great Arc that would give Lambton the true figure of the Earth. When he

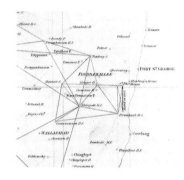

The beginning of Lambton's Great Trigonometrical Survey. His baseline in 1802 near Fort St George is marked on the right. From there he and his team set out to net the entire subcontinent with triangles. By 1806 they had traversed the southern peninsula from east to west and proved it to be 360 miles (580 km) wide on parallel 13°, not 400 miles (640 km) as the Company thought.

One of the Great Trigonometrical Survey's many towers, which still dot India. This one at Radhappuram in Tamil Nadu, south India, was built by Lambton in the first decade of the nineteenth century to provide a vantage point in flat country from which to take observations with his theodolite, and also to serve as a landmark. The needs of the Survey were in constant tension with local people; some rich landowners imagined the Survey was using its telescopes to spy on their women!

died twenty years later something over half the Arc had been measured: a formidable achievement. Lambton's determination may be gauged from his reaction to an accident that struck his beloved theodolite in 1808, at Tanjore, famous for the antiquity, size and significance of its temple. He was on his way to Cape Comorin, and was facing a problem that bedevilled the Survey later in the north Indian plains: how to cope in flat country where buildings and palm trees obscured the line of sight. Lambton's solution was to manhandle the Great Theodolite right to the top of the high temple gateway. Halfway up, a rope snapped; the theodolite's limb crashed against the sculptured gateway and was utterly distorted. "Any person but my predecessor," wrote Sir George Everest, "would have given the matter up as desperate; but Colonel Lambton was not a man to be overawed by trifles, or to yield up his point in hopeless despondency without a struggle." He and the theodolite headed straight for the nearest establishment of ordnance artificers, where Lambton closeted himself in a tent, "into which no person was allowed to penetrate save the head artificers." Six weeks later, by means of wedges, screws and pulleys, Lambton had drawn the limb back almost to its correct circular form, having beaten the bent radii to their proper shape and length with small wooden hammers. The Great Theodolite, though never again perfect, remained in use until 1830, when Everest brought a new design from Britain; after that it was cannibalized to make a second instrument.

From 1815 Lambton rested, exchanging practical observation for elaborate computation. His value for the figure of the Earth excited much interest in European scientific circles. The great French geodesist Delambre wrote to Lambton: "We may flatter ourselves that we know the general figure of the Earth. Let us multiply as much as possible our observations and those scientific enterprises which, like yours, will confirm the glory of the philosophers of the nineteenth century." Lambton was elected a member of the French Académie and then, belatedly, Fellow of the Royal Society. In the face of indifference, he had accomplished his "desideratum more sublime."

The first part of Lambton's proposal—to cover India in triangles—remained incomplete. That same year, 1817, the government took direct charge of the Survey and designated it the Great Trigonometrical Survey (GTS), since it was about to move north into territory outside the control of the Madras Presidency. The Colonel himself, now well over sixty and sick with tuberculosis, made only one further series of observations, in order to instruct George Everest, his new chief assistant. Everest described the occasion he first saw this "great and extraordinary man" at work: "When he aroused himself for the purpose of adjusting the Great Theodolite, he seemed like Ulysses shaking off his rags; his native energy appeared to rise superior to all infirmities; his eyes shone with lustre, his limbs moved with the vigor, of full manhood."

If the courteous, slightly otherworldly Lambton was the founder of the Great Trigonometrical Survey, the intemperate, driven Everest was its engineer. His name is the one that flies from the top of the world's highest mountain; and he is unquestionably the most eloquent chronicler of the Survey that measured it. For twenty years, from the death of Lambton until his retirement in 1843, George Everest *was* the GTS.

In June 1819, following his lessons from Lambton, the twenty-nine-year-old Everest set forth on his first major survey. It was a wilderness northeast of Hyderabad between the rivers Krishna and Godaveri, *terra incognita*, as he termed it, before launching into a vivid account of his youthful approach: "To the north and west and north-east, there were peaks to furnish well-proportioned triangles, and so isolated that there was no doubt of their being reciprocally visible or easily discovered. Further, there were a multitude of small eminences in the neighbourhood, admirably adapted for laying down the whole course of the river; there were also islands in the channel, and an evident facility of fixing the point of conflux of the Pranheeta, a large river which flows into the Godavery.... Three parties were immediately despatched to occupy the three peaks, and I hoped in a few days to complete the observations in which, had success attended me, I should, to use Colonel Lambton's words, 'have performed a very magnificent work indeed to start with.'"

Everest had reckoned without jungle fever—or rather he had chosen to ignore the warnings he had been given. As usual in the Survey, he was working in the rainy season, when the sky is clear of obscuring dust but the country is rife with malaria and typhus. Within days the majority of his team, some 150 men, were "laid prostrate." There was nothing for it but to stumble back to Hyderabad. Fifteen men died and were left beside the road. The rest, when they reached Hyderabad, "bore little resemblance to living beings, but seemed like a crowd of corpses recently torn from the grave."

Six months later Everest tried again. Once more he succumbed to fever. This time he was sent away to the Cape of Good Hope for two years. Although he returned invigorated he never fully recovered: the next year he was so weak that he had the "unpleasant necessity" of being lifted in and out of his seat at the Great Theodolite and having his left arm supported so that he could reach the screw of its vertical circle. Doctors told him to go to the coast if he valued his life, but Everest was adamant: it was a "now or never question... whether the Great Arc should be carried through to Hindoostan, or terminate ingloriously in the valley of Berar."

The perils of his position forced Everest to be inventive, and he now devised methods of observation at night. This was a complete break with Lambton's methods, but Everest did not hesitate. For the Survey, night observation meant the capacity to work in the dry and cold seasons; about three quarters of the year, and rest only during the lethal rains. For nine months, "dry mist" at night was pervious to lights, unlike the wet mist of the rainy season. There were other advantages too. Refraction (the bending of light as it passes through air layers of different temperature and density) was greater and more regular at night. A surveyor could therefore see flares in positions that were hidden during the day. And local people, so vital to survey parties in both sickness and health, were more likely to be free from working their fields outside the rains.

The new seasonal rhythm became the norm for Everest, but meant that neither he nor his long-suffering assistants got much sleep at night. During the day they would be searching the district for sites from which to "blaze away" later. Come nightfall Everest would be scanning the distant darkness for two bonfires burning about 20 ft (6 m) apart: the agreed sign of a suitable station. Then would follow a lot of squinting through the theodolite and finally the ignition of many flares. These came in several varieties. For distances up to about 25 miles (40 km) a vase light was used, consisting of a small cup, 6 in (15 cm) in diameter, filled with cotton seeds steeped in oil and resin, set under a large inverted earthen vessel with an aperture in its side. For distances nearer 50–60 miles (80–100 km) there was the blue light, which was composed of sulphur, nitre, sulphuret of arsenic, camphor, indigo, sulphuret of mercury, gum benzolin, all wrapped in paper, coarse cotton and a sheep's bladder.

Left *Part of the eleventh-century Great Temple of Tanjore in south India. To Lambton it looked like a good observation point for the Survey, so in 1808 he had his beloved theodolite hauled up one of the temple gateways. Halfway up, a rope snapped, the Great Theodolite crashed against the gateway, and its limb was utterly distorted. But Lambton refused to abandon it; with infinite care he beat the limb back into shape and the theodolite continued in service until 1866.*

Left *The Great Theodolite, which now rests at the headquarters of the Survey of India in Dehra Dun. Constructed by William Cary in London, it was sent by sea to India, intercepted en route by a French frigate and taken to Mauritius, then subsequently forwarded to Madras with a complimentary letter. The theodolite weighs half a ton (508 kg) and has a 36-inch (92-cm) azimuth circle. The portraits behind it show Lambton (left) and Everest (right).*

141

"Great care should be taken," warned Everest, 'to prevent the matter exploding in the shape of stars which, however desirable in fireworks, are extremely inconvenient for observation." Burning a blue light sounds almost as risky as catching jungle fever.

That familiar scourge tracked Everest relentlessly along the Great Arc, until in 1825, when the Survey reached an obscure spot called Sironj, even Everest felt ready for a break. He stayed away in England for five years, devoting most of them to boosting the Great Trigonometrical Survey. First he wrote up his recent calculations and published them under the soberly scientific title "An Account of the Measurement of an Arc of the Meridian between the Parallels of 18°3′ and 24°7′, being a continuation of the The Grand Meridional Arc of India, as detailed by the late Lieut.-Col. Lambton in the Volumes of the Asiatic Society of Calcutta." This secured him a Fellowship of the Royal Society. He also cultivated the Directors of the East India Company, advertising the personality behind the copious and baffling GTS reports submitted to England under his name. This would stand him in good stead, he hoped, in the battles between accuracy and economy to come on his return to India.

Everest also travelled, meeting scientists in Europe and spending several months with the Ordnance Survey in Ireland. And he supervised—one can imagine how exactingly—the construction of a new theodolite and "compensation bars" made of iron and brass bound together. These were the invention of the colonel in charge of the Irish Survey, designed to eliminate chains, which expand and contract substantially with temperature. Everest tested them on Lord's Cricket Ground and invited the Directors along. (Few came.)

Back in India, the Company had just appointed Everest both Superintendent of the GTS and Surveyor General of India. All that had gone before had been a preparation for his grand period. Besides devising new methods to complete the Great Arc he would begin the gridiron of triangles that by his death in 1866 would cover most of northern India. One long side of this framework—from Sironj to Calcutta—had been laid down in his absence by an assistant. When Everest reached Calcutta in late 1830 the end of this 600-mile (970-km) longitudinal series of triangles lay not far short of the city. This called for the laying of a new baseline as a check against the length of the final triangular side computed from Sironj. Here was Everest's first real chance to show off his new compensation bars. He selected a site on a road north of Calcutta and ordered two observation towers built that would be visible from the last triangle in the series. Then he called the interested citizens of Calcutta to watch the Survey in action. An elegant breakfast was spread out under tents, for consumption after the ceremonies. The Secretary of the Asiatic Society found himself "contemplating with admiration the order and precision with which the whole process was conducted," while making an atmospheric drawing of the scene.

The new bars performed excellently, in fact almost too well. To Everest's dissatisfaction they showed a difference of 7 ft 11 in (2.4 m) between computed and measured lengths, and a discrepancy of 200 ft (61 m) in height. For a professional craving precision this was serious, but Everest was unable to persuade the Company, either in London or Calcutta, that it mattered. Frustrated, he vowed to make the next stage of the Survey the acme of accuracy.

He decided to divide the Great Arc into two, not counting Lambton's section. The first section would involve resurveying of what he had done from 1819 with inferior instruments, and would terminate in a new baseline at Sironj. The second

Sir George Everest, from a photograph taken in London, probably around 1860. Everest began assisting Lambton in 1819 and took over the Survey on his death in 1821. His grand period, from 1830 until his retirement in 1843, saw the completion of the Great Meridional Arc, which was the backbone of the Survey, running from Cape Comorin to the Himalayas. Everest was ill throughout, sometimes at death's door, but he drove himself and his assistants – British and Indian alike – relentlessly and commanded great loyalty from them, despite a famous temper.

Right The measurement of the Calcutta baseline in January 1832, from a sketch by the Secretary of the Asiatic Society of Bengal, James Prinsep. This was Everest's chance to show off his surveying equipment recently arrived from London. His "compensation bars" are laid out beneath an awning. Above can be seen a specially constructed observation tower; in the foreground lies a chain. An elegant breakfast was later eaten by the interested citizens of Calcutta.

A survey team equipped with elephants sets out from Dehra Dun, where Everest established a new baseline for triangulating towards the Himalayas. Dehra Dun became, and remains to this day, the headquarters of the Survey of India.

would carry the Arc over the plains past Delhi to a baseline in the foothills of the Himalayas. When the process was complete, Everest would compare this baseline in the Himalayas with its length computed from Sironj and find, he hoped, a much smaller discrepancy than in Calcutta. In the meantime his assistants would begin constructing the gridiron by triangulating north towards the Himalayas from the Sironj-Calcutta series. Compelled by this vision, Everest headed for the hills. In the beautiful valley of Dehra Dun, 2000 ft (610 m) above sea level, he found a site for his baseline. Later, the headquarters of the Survey would be established at Dehra Dun, where they still are today.

Everest now faced the most awkward challenge of his career: how to carry the Arc over several hundred miles of thickly wooded and populous level lands between Sironj and Dehra Dun. Lambton had solved a similar problem by hoisting his theodolite to the top of temple gateways. Everest went for specially constructed towers up to 60 ft (18 m) tall as the only viable solution. Of the fourteen built for him, thirteen still stand firm, relics of a bygone age.

Before the towers could go up, the best positions for them had to be chosen, to ensure that they were visible to each other. Selection was tricky, since observations had to be performed at night to avoid the dusty haze of the plains. Groves of mango and tamarind trees, ancient convoluted banyans, temples, mosques and other buildings all stood in the way. How was a surveyor to know where a potential advance station was if he could not see rays coming from it? After repeated failures to make contact, Everest invented ray-tracing to identify the path of a hypothetical ray and decide if it might be worth clearing that path of obstacles. Ray-tracing worked in two ways: one slow and accurate, the other convenient but sloppy. In the first case a series of minor triangles was extended in the general direction of the invisible station until its bearing was located: this gave Everest a clear picture of everything lying between. The second method was simply the oldest method of Indian survey, used since Rennell's first route surveys seventy years before: the perambulator traverse, in which a large wheel was pushed by a surveyor to measure distance like a cyclometer on a bicycle. As with triangulation, the surveyor set off blindly in the expected direction of the station and took bearings as he went, until he sighted it.

Everest took a robust attitude to anything, animate or inanimate, that stood in the way of the Survey. "Noble trees" he tried to respect, religious sensibilities he bore willy-nilly, but British political officers and native potentates he scorned if they proved uncooperative. When a wealthy Indian landowner objected that Everest's towers would violate the privacy of his womenfolk, a common complaint against the Survey, Everest mocked: "The persons of the Great Trigonometrical Survey are of too good taste to concern themselves with Zalim Singh's zenana. He does not do them justice. Persuaded that our telescopes which invert have magic powers, and are able to turn women upside down (an indecent posture no doubt and very shocking to contemplate), it is natural enough that they should assign to us the propensity of all day long spying through the stone walls at those whom they deem so enchanting."

In 1838 he turned some equally elaborate sarcasm on his esteemed colleagues of the Royal Society. Behind his back, thirty-eight Fellows, some of them eminent, had signed a petition to the Company advancing the cause of a certain Major Jervis to take over the GTS, since Everest himself seemed too ill to continue. In a series of trenchant public letters to the Duke of Sussex, the Society's President and, Everest reminded him, "a brother Freemason," Everest demolished their knowledge of the Survey and remarked that they would shortly resemble a character in a famous scene from Shakespeare: "Oh Bottom, thou art changed! What do I see on thee?" The blast worked: the Fellows were not prepared to risk looking like asses. No more was heard of the petition or of the presumptuous major.

Everest was incapable of operating at any pitch other than a high one; and yet his assistants, European and Indian, seldom left him. Like Lambton, he inspired an exceptional degree of loyalty, probably because both men treated everyone without prejudice, Indian assistants included: "Without provision for themselves in case of their being crippled by sickness, accident, or age, or for their families in the event of their death, they yet ventured fearlessly and without a murmur to face those awful dangers which would have made the stoutest hearts quail and shrink," Everest wrote of Lambton's Indian assistants (and by implication his own). "There is a forlorn and desolate feeling produced by the thought of yielding up one's life in a wilderness, with none but jackals to sing our requiem, and tigers to prowl, and vultures to flit round our tombless corpses, which harrows up the soul with inexpressible horror." Everest ought to have known, having lain at death's door on Survey more than once. In 1840, when he was forty-nine, the old war-horse admitted that "age has begun to leave the usual indelible marks upon me. My eyesight, once so vigorous, is failing me fast. I have habitual attacks of gout and indigestion. My native energy and activity are forsaking me. It is indeed better that I should leave fieldwork for younger men. . . ." But not, Everest said, until the Great Arc was done.

And so, storming and raging, praising and scolding, and sometimes patiently explaining almost twenty-four hours a day, some ten years after his return from Britain George Everest brought his team and triangles to his baseline at Dehra Dun, within view of the awesome Himalayas. Surprisingly—or perhaps it is typical of Everest's overriding obsession with the Survey—he does not record his feelings on seeing the distant peaks. He was interested in the discrepancy between his triangles and his baseline. It measured 7 in (18 cm) over about 500 miles (800 km): a well-deserved triumph.

Everest and his chief assistant Andrew Waugh now began the series of astronomical observations at either end of the sections of the Great Arc that would both generate the most precise figure of the Earth yet calculated and lead to a discovery whose scientific significance remains puzzling today. The elevations of the same group of stars were measured from two small observatories on the 78° meridian near Sironj and Dehra Dun, at exactly the same time. Latitudes were calculated, along with the distances between the two observatories. The astronomical survey placed them about 100 ft (30 m) closer together than the triangulation—obviously an enormous discrepancy by the standards of Everest.

He had a suspicion why this was so, and for once it made him accept a discrepancy almost blithely. Both he and Lambton had anticipated that near the vast mass of the Himalayas gravity would vary from its normal value. They expected the difference to show in their measurements of stars, because the zenith sectors were aligned with plumb lines: the Himalayas would attract the plumb line slightly away from the vertical and so distort the reading of the zenith. Neither pioneer felt equal to the task of calculating how much distortion there would be. "No territorial measurement has yet been free from [these errors] and probably none ever will be," wrote Everest.

It remained for a scientifically minded archdeacon of Calcutta, John Henry Pratt, and a distinguished Astronomer Royal, George B. Airy, to start the argument over how to interpret the discrepancy of the Survey. They estimated that it should be much greater than it actually was, unless one assumed that the Himalayas were resting on much less massive foundations than predicted by the average strength of the Earth's surface. What really holds up the Himalayas is a live issue well over a century later. There is a missing mass below the mountains that needs explaining.

Pratt and Airy agreed that the Earth's crust, of which the Himalayas form part, was floating on a denser mantle, like cream on milk. (Geodesists refer to this phenomenon as "isostasy.") Pratt suggested that the thickness of the crust would be the same everywhere, but that crustal temperature and density might vary from place to place. Where the crust was hotter and lighter than average it would rise to form mountains. Airy disagreed. He viewed the crust as uniform in density, but variable in thickness. Where mountains occur, the crust is thicker; he compared peaks to the tips of icebergs, each having deep, invisible roots. Although the two theories predict different crustal thicknesses, both assume a smaller-than-average mass of rock below mountains.

Seismological studies in this century show that Airy's iceberg theory is the right one, though it requires modification. The crust below India is 20–25 miles (32–40 km) thick, increasing below the Himalayas to 35 miles (56 km). But if the Himalayas were supported by deep crustal roots alone, as Airy's theory states, this depth should be as much as 50 miles (80 km). The actual figure is 40–45 miles (64–72 km), *less* than the thickness of crust below the Tibetan plateau north of the Himalayas, and considerably less than the depth of crust below the Andes. Why this apparent contradiction of common sense? Because the Tibetan plateau and the Andes are supported mainly by deep crustal *roots*, while the Himalayas seem to be resting on a crustal *plate* that is being pushed upwards by the pressure of continental drift that is forcing India steadily northwards into Asia. Quite how this works is not known since, like George Everest himself, we still have insufficient knowledge of gravity variations in the Himalayas. Until we get these, probably from satellites, those "fascinated by mountain architecture," in the words of a US geophysicist, "will remain in a situation not unlike that of Gothic architects, who found they could support giant cathedrals with flying buttresses but who never really understood the underlying physical principles."

The Great Trigonometrical Survey did not begin triangulating the remote snow-capped summits from dusty Hindustan until after George Everest's retirement in 1843. The job fell to Captain Waugh, his assistant and worthy successor, who supervised a group of surveyors observing the high peaks between 1845 and 1850. The North East Longitudinal Series, the last link in the gridiron, was probably the most lethal of any survey in the history of the GTS. In one season forty Indian assistants died of fever; in another the whole surveying party was conveyed back to base in a helpless condition. When a new officer took charge, he too went down. "He hurried towards Darjeeling," the *Historical Records* of the Survey state, "and was found dead in his dooly when it arrived." Of the five officers active on this series, two retired and two died: victims of the destructive climate.

The snowy wastes were not responsible, perhaps contrary to expectation. Indeed lives would have been saved if the GTS had been permitted to get closer to the peaks. But the Gurkha government in Nepal had banned British visitors, so the surveyors were compelled to operate in the *terai*, a long swathe of low jungle along the border. They had to observe the elusive peaks from distances of up to 150 miles (240 km). Accuracy inevitably suffered, especially given the extra hurdles of weather, unpredictable refraction and deflection of the plumb line. These were the reasons why earlier surveyors had not been able to verify their suspicions that the Himalayas

Right *Mount Everest, or Peak XV as it was first designated, and Radhanath Sikdar* (below), *Chief Computer of the Great Trigonometrical Survey of India. Sikdar was a gifted Bengali mathematician and the first educated Indian to join the Survey, recruited by Everest in the 1830s. He may also have been the first to realize the Survey had measured the highest mountain in the world.*

were the highest mountains in the world—a prediction first made as early as 1784 by Sir William Jones, the brilliant founder of the Asiatic Society, after he had gazed upon distant peaks from the plains of Bengal.

European geographers had remained sceptical of claims such as that of Jones, preferring to remain loyal to the Andes. Hence the reluctance of Waugh, the Surveyor General, to make a quick announcement to the world that the Survey had found the supreme summit. Bengalis believe that around 1852 the survey's Chief Computer Radhanath Sikdar, a gifted Bengali mathematician, rushed into Waugh's office and declared, "Sir, I have found the highest mountain in the world!" The supposition is a reasonable one, but there seems to be no hard evidence for it. Whatever the truth, the official statement was not made until mid-1856 at a meeting of the Asiatic Society in Calcutta (endorsed by Sikdar).

Peak XV, as the highest peak was known, had been observed, along with many other peaks, at several times by different surveyors. In 1850 it was triangulated from six stations in the plains up to 120 miles (190 km) away, and six heights were computed: 28,992 ft (8837 m), 29,005 ft (8841 m), 29,002 ft (8840 m), 28,999 ft (8839 m), 29,026 ft (8847 m) and 28,990 ft (8836 m). The average, 29,002 ft (8840 m) was adopted. This remained the official height despite complex debate on where to fix the figure of the Earth beneath the mountain and hence its true height, until 1952–4, when a party of Indian surveyors went into Nepal and measured the mountain at a distance of only 50 miles (80 km). The height is now taken to be 29,028 ft (8848 m).

The choice of a name was controversial too. The Survey had a policy, inaugurated by George Everest, of calling natural features by their local names. Ironically Mount Everest is the only Himalayan giant with an exotic English name.

The celebrated Pundits – Indians trained by the Survey of India – entered forbidden Nepal and Tibet in the years after 1865 to take surreptitious measurements and observations. Some were disguised as Buddhist lamas. Instead of carrying the traditional 108-bead rosary, they carried 100 beads and used them to mark off their paces – exactly two thousand steps to the mile (1.6 km). This rate was drilled into them before they set out on their hazardous journeys.

Waugh took the decision after searching in vain for a local alternative. As soon as the name was made public in 1856, the former British Political Resident at Katmandu challenged Waugh, suggesting a Nepali name, Devadhanga. But his identification was vague, totally inadequate to satisfy the rigorous Survey. A further suggestion, Gaurisankar, was disproved in 1903 when the Survey was briefly allowed into Nepal: Gaurisankar was another peak some 35 miles (56 km) from Everest. A Tibetan name Qomolangma describes the entire massif of which Peak XV is the summit; it appears on some modern maps as an alternative name for the summit itself.

Exciting as it was, the measurement of Mount Everest by no means signalled the end of the Survey, or of the astounding achievements of its unsung heroes. Virgin territory remained in the deserts of western India, in the jungles of Assam, and in the remote northwest; by 1880 it would all be basically triangulated. The most extraordinary of these surveys was the one in Kashmir, where by 1862, thirty-seven peaks over 20,000 ft (6096 m) had been scaled with a theodolite. For twenty years the world's altitude record was held by one of the Indian assistants who carried a signal pole to 23,050 ft (7026 m), yet we do not even know his name. When a British surveyor returned to one of these peaks in 1911 he found the raised survey platform and finely chiselled markstone firmly in position. Nearby was a ruined stone shelter, in the corner of which lay a human skeleton. No wonder Everest had written of "a forlorn and desolate feeling... which harrows up the soul with inexpressible horror" when contemplating the courage of the Survey's Indian assistants.

Two years after this visit northern Kashmir was the site of a historic encounter. In August 1913 the triangulations of India and the Russian territories in Asia were officially joined "in the interests of science" when Indian and Russian survey parties met. They compared the length of their junction side: 23,413.5 ft (7134.9 m) for the Russians, 23,408.5 ft (7133.4 m) for the British. This marked the official end of a curious period in the recent history of central Asia, known as the Great Game, in which the Survey had had a chance to play its most romantic role yet.

Rudyard Kipling dramatized the Great Game in *Kim*, published in 1901, his famous story of a street-wise orphan trained by Colonel Creighton to explore the Himalayas as a spy with the help of a loquacious Indian guide, the mysterious Bengali Babu. Something similar had actually happened in India from the mid-1860s. By 1865 the British were standing at the gateways to Nepal and Tibet without hope of entry into the Himalayan fastness. The two states were blanks on the map. Nepal was officially barred to them by treaty, while in Tibet the Chinese Emperor had long declared that all *feringhees*—foreigners—were unwelcome. Anyone who got through in disguise might well be beheaded if spotted. And so the indefatigable officers of the Survey trained Indians to go where they could not.

These intrepid explorers—more than fifteen in all—became known as the Pundits, or teachers, after the best known of them, Nain Singh, who was a headmaster in his village near the border with Tibet. He and his three cousins became the most celebrated of the Pundits. For over fifteen years in various guises they tramped the immense bare plateau of Tibet where no Englishman and almost no European had gone before, eventually returning in tatters to Dehra Dun to report to their masters what they had seen.

Before they went forth they had to be trained. The Survey intended them to take surreptitious measurements of distance, latitude and height. The favorite disguise was that of a Buddhist lama, or priest. A sergeant-major drilled Nain Singh with his pace-stick until he could walk at a precise pace—exactly two thousand steps to the mile (1.6 km). A hundred-bead rosary (instead of the usual 108 beads) allowed him to count his paces, and inside his prayer-wheel constructed in the Survey workshops, were compasses for taking bearings while at prayer. In the false bottom of his travelling chest was a sextant, and inside sealed cowrie shells mercury for his artificial horizon. Nain Singh's salary was a paltry 20 rupees a month.

The journeys of the Pundits are the stuff of epic poetry. Unfortunately, they were not writers. Yet from the accounts they noted down or dictated to the interpreters of the eagerly awaiting Captain Montgomerie, we can marvel at their endurance and get vivid flashes of the perils that beset them. During Nain Singh's visit to Lhasa in 1866, for example, he was obliged to pay his respects to the Dalai Lama, the God King, in his throne room at the Potala. Although Nain Singh was not devout, his ancestors had been. "You can imagine his feelings when ushered into the great Lama's presence with his prayer-wheel stuffed with survey notes and an English compass in his sleeve," Montgomerie wrote to the President of the Royal Geographical Society in London. The thirteen-year-old boy-god did not use the psychic powers with which he was reputedly gifted.

Kinthup, an illiterate Sikkimese from Darjeeling, was perhaps the most devoted of all the Indian explorers, and for many years the least sung. He was originally despatched as a servant to a Mongolian lama, who had instructions to enter Tibet and follow the course of the Tsangpo river east as far as he could go. There the lama was supposed to cut five hundred logs to a prescribed length, drill a hole in each, insert a short metal tube and then heave the logs into the Tsangpo at the rate of fifty a day over a prearranged period. Downstream in the foothills in India, Captain Harman of the Survey would be waiting for them. This apparent charade was designed to solve a mystery that had exercised the best minds in geography since before Rennell: were the Tsangpo of Tibet and the mighty Brahmaputra of Bengal one and the same river?

Four years later Kinthup returned alone to Darjeeling. Harman was dead and no one in authority knew of the fantastic mission he had sent out. The lama had dawdled along the route and eventually vanished after selling Kinthup into slavery. Seven months later Kinthup managed to escape and decided to complete the lama's task. He entered a monastery, pursued by servants of his slavemaster. The head abbot took pity on him and bought him for 50 rupees. After four months he got leave to go on pilgrimage. Instead he cut five hundred logs and hid them in a cave. Then he returned. After some more time Kinthup got permission to go on pilgrimage to Lhasa. Instead he used his visit to dictate a letter to the Survey in Darjeeling. Then he returned to the monastery for nine more months. Again he asked to go on pilgrimage. This time the head abbot told him he was free. He went back to the cave and launched the logs on the days mentioned in his letter. Down in the plains no one saw them pass, such was the time that had gone by since Kinthup had left. His precious letter had gone astray.

A hundred years and a world removed, it is not easy to comprehend what made these men willing to risk life and health for the Survey of the Himalayas. Certainly they did not do it for money. Nain Singh, on his retirement from the Survey, received a village in the plains and the highest award of the Royal Geographical Society; but other Pundits earned only small pensions and Kinthup got only 1000 rupees, thirty years late and just before he died, from a reluctant government. The motivation of the hundreds of surveyors and their assistants who mapped the subcontinent down below for a century or more is nearly as hard to fathom.

*Daniel Boone,
frontiersman and
surveyor, painted by
Robert Lindeux. Between
1767 and 1773 Boone
explored Kentucky and
later helped to colonize it,
going fifty-fifty with
settlers on all the land he
surveyed. He formed
the first settlement
in Kentucky at
Boonesborough, which he
courageously defended
against Indian attack.
Boone died in 1820 in
Missouri, where he had
been driven by disputes
over land.*

9

OPENING UP AMERICA

As the Great Trigonometrical Survey of India drew to a close, mapping of the emerging United States—with a total area nearly two and a half times that of India—was getting into high gear. From the outset the driving force was different. In India maps were made by a foreign government to control a subcontinent densely settled; in America maps were a crucial aid to white settlement as it spread west, displacing Indians (Native Americans) from the promised land. The mapping of India was centrally directed; that of America pragmatic and piecemeal. The first calls to mind the famous "steel frame" of the Indian Civil Service, the second the patchwork quilt of the early colonists.

Maps of America were wanted quickly, and needed above all to be practical. There was no need to show every detail: settlers thinking of moving west wanted to know how to get there and what hazards they would face en route. Once there they were interested in the lie of the land if they were planning to farm, the location of minerals if they were prospectors, or the routes of rivers, roads and railroads if they were going into business. They were not particularly concerned to know their precise latitude or longitude in relation to the government back east in Washington DC.

The requirements of the time discouraged the American Congress from launching a national trigonometrical survey, but they did not prevent frequent clashes between the states and the center, between the man on the spot with an urge to get rich and the scientist-cartographer with the whole nation in mind. In 1856 Senator John B. Weller of California presented to Congress a leather-bound petition signed by 75,000 California freemen calling for a railroad over the Sierra Nevada. Do not entrust the project to engineer-explorers, he demanded, but to "practical men—stage contractors, who, instead of taking instruments to ascertain the altitude of mountains take their shovels and spades and go to work and overcome the difficulties of the mountain, while an engineer, perhaps, is surveying the altitude of a neighbouring hill."

Similar views had been held for more than a century, since before the American Revolution. On the east coast, where surveying began, it was a colonial commonplace that what worked in Europe would not necessarily work there. The surveyor's standard textbook recommended that the theodolite should not be employed in

George Washington, by James Peale the Elder. Washington began life as a surveyor in Virginia and continued surveying throughout his life, lamenting as president the poor quality of maps available to the new United States. This is the only portrait of him executed before the Revolution.

America, being more suited to the cleared lands of Europe. Notes accompanying the *General Map of the Middle British Colonies in America*, published in 1755, observe that "everywhere [is] covered with woods. . . . Here are no churches, Towers, Houses, or peaked Mountains to be seen from afar, no Means of obtaining the Bearings or Distances of Places, but by the Compass, and actual Mensuration with the Chain." Surveying at that time was a lucrative business and a recognized means of social advancement. For a county surveyor in Virginia the remuneration was 40 lb (18 kg) of tobacco for every 100 acres (40 hectares) surveyed. On the frontier itself the surveyor was generally paid in land.

Virginia-born George Washington began his career as a surveyor. In 1745, at the age of thirteen, he borrowed some of his father's instruments—a brass surveying compass, a Jacob's staff, a chain and rods—and began running lines at home and on adjacent plantations. Three years later he rode with a friend, the eldest son of Lord Fairfax, and a veteran surveyor, across the Blue Ridge into the "Valley of Virginia," then sparsely settled. Before the trip was over, Washington had chosen the career he would follow as a young man. The next year he was sworn in as the surveyor of Culpeper County. More than a hundred maps surveyed or annotated by Washington exist from his time there. He proved thorough and accurate at the work, but not an outstanding surveyor.

After five years, when he was twenty-one, Washington was commissioned by the

The Blue Ridge Mountains, across which George Washington rode in 1748, aged sixteen, into the Valley of Virginia, then the frontier of America. This trip with two friends, one of them a veteran surveyor, persuaded Washington to take up surveying as a career. The following year Washington was sworn in as surveyor of Culpeper County.

Virginia militia, and for a decade his work as a surveyor was interrupted. But on resigning his commission in the 1760s, he travelled extensively and made substantial purchases of land, mapping them himself with instruments ordered from London. Even as a general in the Revolution he continued the practice, constantly lamenting the poor quality of topographical mapping in the United States. On one occasion, he informed the Congress: "I have in vain endeavoured to procure accurate maps . . . but was obliged to make shift with such sketches as I could trace out from my own observations and that of the Gentlemen around me." At the time of his death, in December 1799, Washington still possessed a full set of surveying instruments.

The famous survey of Charles Mason and Jeremiah Dixon was the first in America to go further than a property survey. It started in 1764, the same year as Rennell's Bengal Survey. The Mason-Dixon Line later acquired a heavy significance as the division between North and South—freemen and slaves—during the 1861–65 Civil War; but it began as the solution to a long-running dispute between two powerful families. This conflict went back to 1632 when George Calvert, the first Lord Baltimore, had received the lands of Maryland from the British Crown, as far north as the "Fortieth degree of North Latitude." Subsequently William Penn had been granted what would become Pennsylvania, also by the Crown. The problem was that the northern boundary of Calvert's land appeared to run through the center of Philadelphia.

In 1761 the Penn and Calvert families jointly appealed to the Astronomer Royal to send someone mathematical from England to fix a proper boundary without bias to either side. By the time Mason and Dixon arrived in the New World, fresh from observing a transit of Venus at the Cape of Good Hope, the Penns and Calverts had agreed to a boundary 15 miles (24 km) south of the southernmost limit of Philadelphia. They asked Mason and Dixon to define this line exactly. First the surveyors needed to define the edge of Philadelphia, by sixty observations of stars spread over six weeks, which yielded the latitude 39° 56′ 29.1″—only 2.5′ off its true position. A point 15 miles south of this would lie in New Jersey, so Mason and Dixon had to move about 30 miles (48 km) due west to begin their line. There they established what they afterwards called "the Post mark'd West."

The survey proper began in spring 1765, after they had surveyed the north-south boundary between Maryland and Delaware. The work was tough and hazardous, the way blocked by forests and deep snow, and they faced extreme cold, "border ruffians" and Indians. Up front were the axemen, clearing a corridor about 30 ft (9 m) wide. Every ten minutes in a degree—that is about every 11 miles (18 km)—a halt would be called to check the survey's position by astronomical observations. By October they had crossed the Susquehanna river and reached the summit of the Blue Ridge. Then the team went back east for the winter. The following spring was spent replacing temporary wooden posts with stone markers; summer and autumn in extending the line until it reached five degrees of longitude west of the start-post, as stipulated by the Pennsylvania charter.

Further progress required negotiation between colonial officials and Indians. Mason and Dixon were given an Indian escort and warned that they should treat them and other Indians well because "the public Peace and your own Security may greatly depend on [their] good Usage and kind Treatment." The work was slower than before. With persistent rumors of trouble from Shawnee war parties, most of the Indian escort faded away. The axemen threatened to depart. On 9 October 1767 the survey finally came to a standstill, 30 miles (48 km) east of what is now Pennsylvania's south-west corner, at a creek the axemen refused to cross. The few remaining Indians in the party said: "You go no further." And since Mason and Dixon knew that a small township at this creek had been torched by Indians twelve years earlier, they were not disposed to disagree. Their final observations were taken "233 Miles, 3 Chains and 38 Links [roughly 375 km] from the Post mark'd West," 40 miles (64 km) due south of Pittsburgh.

From that town was launched the second path-breaking survey in America, nearly twenty years later. When Thomas Hutchins reached it in September 1785, Pittsburgh was more a rough-and-ready village than a town; a group of log cabins with three hundred inhabitants, on the threshold of the American West. Hutchins already knew the region well, having served there as a military officer before the Revolution and led expeditions down the Ohio river to the Mississippi and overland to lakes Erie and Michigan. During the Revolution he had become a surveyor on the southern front, and eventually geographer to the Continental Congress. Now the Congress required him to divide into lots the lands west of Pittsburgh and north of the Ohio, in order to raise some quick revenue.

The method Hutchins applied determined the future course of US land survey. A glance at the straight lines and right angles of today's state boundaries, or the chessboard layout of the United States from the air, demonstrates clearly how the demands of rectangular survey took precedence over the terrain. This was the way, decided Congress, that would be "attended by the least possible expense, there being only two sides of the square to run in almost all cases," with consequent "exemption from controversy."

Hutchins' survey was directed to begin at a stake on the Ohio river that had been planted by another survey party, who had extended the Mason-Dixon Line to the southwest corner of Pennsylvania before running it due north to the Ohio. From this so-called Place of Beginning, Hutchins was to lay a carefully measured baseline and along it mark off *ranges* that would become townships. Each of these was to be 6 miles (9.7 km) square, containing thirty-six sections of one square mile or 640 acres (259 hectares). Within each range important features, such as "mines, salt springs, salt licks, and mill seats," were to be indicated. Hutchins' goal was thus more than that of a normal county surveyor.

Pittsburgh provided Hutchins with axemen, chain-carriers, horses and supplies. He also expected thirteen state-appointed surveyors, but only eight showed up. The colonel in charge of the fort gave assurances about the local Indians, and Hutchins and his party headed for the stake by the Ohio. But just 4 miles (6.4 km) of baseline had been completed when Hutchins got wind of Indian raids in the vicinity. In the words of a Shawnee chief, "We do not understand measuring out the land—it is all ours. . . . Brothers, you seem to grow proud because you have thrown down the king of England." Immediately Hutchins fell back across the Ohio, and thence to Pittsburgh since no armed escort was available to protect the survey.

Nearly a year later, he returned, with 150 troops and twelve of the thirteen surveyors requested. He made progress for about four months until once more he had to retreat from attack, this time by marauders who drove off all his horses. Only four ranges of townships had been entirely surveyed, while three others lay partly done. The area therefore became known as the Seven Ranges. Unfortunately their boundaries failed to meet cleanly at right angles because Hutchins had not allowed properly for the curvature of the Earth that compels a surveyor to make a baseline at a given latitude slightly curved, not straight, and to allow for the fact that "parallel meridians" actually converge over a long enough north-south distance. Instead of calculating these discrepancies, Hutchins seems to have estimated them both. It was left to one of his assistants, on a later survey, to devise a compromise between the demands of rectangularity and convergency.

Nevertheless in autumn 1787 Congress held a public auction of the surveyed land in New York City. It raised a little over $100,000. A disappointed Congress changed its policy: the West would now be sold off in large tracts to land speculators. Hutchins was invited to take part in the surveying, but he declined. His dreams lay further west, in the valley of the Mississippi, then part of the Spanish empire. He had proposed to the Spanish emissary that he become geographer to the Spanish instead of to the Congress. But within a year he died, aged fifty-nine and still in Pittsburgh, before his offer could be taken up.

The notion of a Mississippi survey was only a little ahead of its time. The eyes of most movers and shakers in the young US nation increasingly turned west in the 1790s. Thomas Jefferson, the first Secretary of State, and from 1800 the President, was especially interested in expansion. In April 1803 he got his opportunity. Napoleon Bonaparte had just forced the Spanish to give back to France the area called Louisiana—all the land between the 49th parallel in the north, the Mississippi in the east, and the ill-defined Stony Mountains in the far west. Now he suddenly decided to sell it to the United States, fearing that otherwise it would be grabbed by the British in the coming war between Britain and France. The price agreed on was $15,000,000,

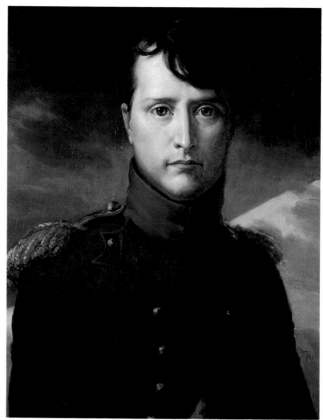

Napoleon Bonaparte as First Consul of France, painted by François-Pascal-Simon Gérard in 1803. That year he sold Louisiana to the United States for $15,000,000 to prevent it being grabbed by Britain in the coming war between France and Britain. The Louisiana Purchase more than doubled the size of the United States.

Thomas Jefferson, painted by Rembrandt Peale in 1800, was the first US statesman to contemplate American expansion to the Pacific coast. The idea became reality in 1803 with the Louisiana Purchase from France, the biggest land sale in history. The following year he despatched Lewis and Clark on their historic journey across America, and gave Americans their first idea of the West.

some four million dollars more than the representative of Congress in Paris had been authorized to pay. But that sum had been intended to buy just the city of New Orleans, not an empire! Here was the biggest land sale in history, at a bargain-basement price.

The Louisiana Purchase more than doubled the size of the United States. The new area was 828,000 square miles (2.14 million sq km): two-thirds the size of India. At one stroke it opened the prospect of expanding the country to the very shores of the Pacific Ocean, where American ships travelling via Cape Horn had already cornered a highly profitable trade in furs from the American Northwest to the Orient. The acquisition gave Americans a new idea of themselves, the idea of the West that would dominate their imaginations for the rest of the century. The West would become, at various periods, sometimes simultaneously, "the great empty continent, Eldorado or Cibola, a barren waste of heathen savages and Spaniards, the passage to India, an imperial frontier, a beaver kingdom, the great American Desert, a land of flocks and herds, a pastoral paradise, an agricultural Arcadia, a military and administrative problem, a bonanza of gold and silver, a safety valve, a haven for saints, a refuge for bad men, and ultimately, towards the end of the nineteenth century, an enormous laboratory," in the words of William H. Goetzmann's prize-winning *Exploration and Empire*.

In 1803 no one had a clear idea of what had been bought from France. Notably mysterious were the extent and height of the mountains that formed the Continental Divide, and the course of the Missouri before it turned south at its Great Bend, finally to flow into the Mississippi. In fact that old myth, the Northwest Passage, was still much in people's minds, the third voyage of Captain Cook notwithstanding. The discovery in the 1790s of the lower reaches of the Columbia river, missed by Cook, had suggested that the river might somehow cut through the Continental Divide and connect with the headwaters of the Missouri, thus permitting water-borne travel right across the West. This seemed reasonable, given that the two rivers appeared to flow at a similar latitude—especially if you believed the mountains to be only a few thousand feet high, as Aaron Arrowsmith had postulated on his 1795 map, based on the sketches of a Hudson's Bay Company surveyor. A second map, the 1797 "New Map of North America Showing all the New Discoveries" by Jedidiah Morse, hopefully shows the River of the West rising near the source of the Missouri. Even if the connection was not direct, it was widely asserted that only a short portage separated the rivers; as little as 20 miles (32 km), said some.

President Jefferson, gazing at such maps in Washington DC, was a convinced advocate of the water route to the Northwest and on to Cathay and India: the same dream that had impelled Columbus three centuries before. If it existed, not only could it halve the average round trip to China to a year and a half, it would allow the Americans to challenge the rival North West and Hudson's Bay Companies for the fur trade of the Northwest. A land expedition to the Pacific coast, long contemplated, now seemed imperative.

To head it Jefferson selected Captains Meriwether Lewis and William Clark, justifiably the most famous names in the exploration of the American West. Lewis was twenty-nine, Clark thirty-three; both had extensive experience in the wilderness, and Lewis had spent two years as Jefferson's private secretary, preparing for his mission. He had meanwhile gained a background in natural history, astronomy, and rudimentary surveying techniques from the best minds of the American Philosophical Society at Philadelphia. Distinguished and melancholy, Lewis was the diplomatic and commercial thinker of the expedition, the bold and outgoing Clark its negotiator, according to Bernard DeVoto, who edited the modern edition of their journals. They "worked together with a mutuality unknown elsewhere in the history of exploration and rare in any kind of human association."

Despite its romantic image, the expedition was highly planned. An estimate of the distance to the Pacific was calculated from the longitudes announced by the various explorers of the Northwest in the 1790s. Jefferson's instructions to Lewis and Clark were lucid and explicit. They were to survey in detail the water route up the Missouri and down the Columbia. They were to investigate a direct overland route between the United States and the Northwest maritime trade. They were to amass information for a US challenge to the North West Company's beaver trade. They were to negotiate with the Indians, impress upon them the advantages of trade with the Americans over trade with the French, Spanish or British, and foster intertribal peace. At all times they were to take observations "with great pains and accuracy," fix their latitude and longitude—astronomically, not by chronometer—and write them down in several places for safety. "A further guard would be that one of these copies be written on the paper of the birch, as less liable to injury from damp than common paper," Jefferson recommended.

The only instruction left implicit was perhaps the most important one of all, which had no doubt been impressed upon Lewis during his spell as Jefferson's secretary. By making this land traverse to the Northwest ahead of the British, the expedition would buttress the United States' claims on the Oregon Territory, following the American traders on the Pacific coast.

On 14 May 1803 the party set out from St Louis in a sturdy keelboat holding forty-five men in total. A year and a half later William Clark carved on a tall pine

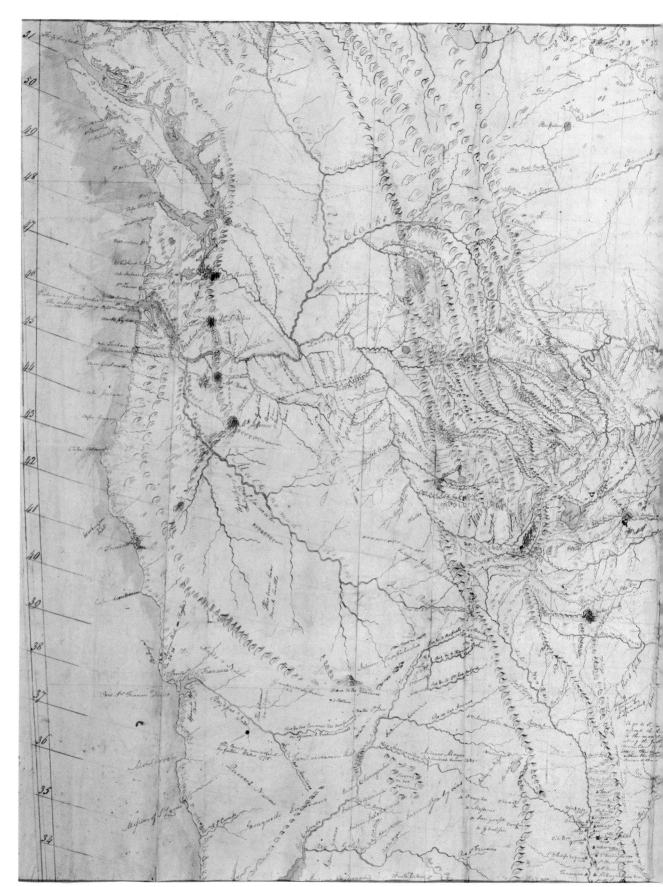

"A map of part of the Continent of North America" by William Clark, 1810, "compiled from the information of the best-informed travellers." The map, kept by Clark in his office in St Louis and regularly updated, is one of the most important maps ever drawn in America. In the Far West, at the entrance to the Columbia river, Clark marked the destination of the Lewis and Clark expedition. The scale of his map was 50 miles to the inch (32 km to the cm).

A MAP 1781
of part of the Continent of
North America

where the Columbia river enters the Pacific: "William Clark December 3rd 1805. By Land from the U. States in 1804 and 1805." When they returned nine months later, they had proved conclusively that the Northwest Passage did not exist, that the Rocky Mountains were a formidable and complex barrier across the West, and that about 220 miles (350 km) separated the source of the Columbia from the Missouri. The hoped-for navigable route was impossible, especially as it would have meant negotiating the Great Falls further down the Missouri. More positively, the expedition had discovered five passes through the Rockies into the Oregon Territory, and brought back a copious array of new knowledge about the West. It had "given not only the Oregon but the entire West to the American people as something with which the mind could deal," comments DeVoto. "Here were not only the Indians but the land itself and its conditions: river systems, valleys, mountain ranges, climates, flora and fauna, and a rich and varied membrane of detail relating them to one another and to familiar experience."

As for the Indians, the expedition was a "bright strand in a dark history." Clark had such a gift for friendship that before his death in 1828, this "Red Headed Chief" was known to all the tribes of the Plains and Northwest. After the expedition he made his home in St Louis, where any Indian delegation to the city would first seek him out. If a fur company was going up-country, it would look for credentials from Clark; if a US embassy was heading for Indian country, it would try to take Clark along. Miracles were expected of him by the Indians; what he could he did, and it was more than any other individual in the nineteenth century.

In mapmaking Clark put this rapport to good use. On the expedition Indians drew maps for him on a hide or on the ground with charcoal or a piece of stick. This

James Bridger, trapper, guide and explorer, typical of the "mountain men" who knew the West decades before it was properly mapped but took most of what they knew with them when they died. In 1825 he became the first white man to see the Great Salt Lake; he was also one of the first to see the Yellowstone region. The rock which bears his name carved on it is in Wyoming, on the Oregon Trail.

information, which was far more accurate and detailed than from any other source, Clark eventually transferred to a master map of the West that he kept in his office in St Louis. Constantly improved through discussion with Indian delegations and with trappers, this map grew into one of the most important maps ever drawn in America. Indians and trappers, the "mountain men," remained the prime source of cartographic knowledge about the West in the first three decades of the nineteenth century, until the start of army exploration. Mountain men such as Jedediah Smith went just about everywhere in the Rockies in pursuit of beaver and other fur. Most of what they knew died with them, but their reticence is understandable since they were operating illegally from the point of view of Spain, which owned much of the Far West until the 1846–48 Mexican War with the US, and since they also wanted to protect their sources of fur.

There were exceptions, more open with their information, such as Peter Skene Ogden, a ruthless agent of the North West Company who once assaulted a Hudson's Bay Company official in his own post and left him at the point of death. Between 1824 and 1830, Ogden led six expeditions, completely exploring the Snake River country, north and south, Oregon, the Great Salt Lake and Bear River region, much of California north of San Francisco Bay, and the Humboldt River. He was also the first person known to have traversed the transmontane West from north to south. Not surprisingly Ogden's accounts were lapped up by the mapmakers Aaron Arrowsmith and Sons in London and A. H. Brué in Paris. Ironically, the maps he helped to create inspired the Americans planning to drive the British fur-traders out of the Northwest.

A more typical mountain man was the self-employed James Bridger. As a young man in 1825, he was sent down the Bear River to settle a bet between two older trappers as to its destination. This led him to discover the Great Salt Lake. After tasting the water he returned, apparently convinced that the lake was an arm of the Pacific. He put nothing in writing, but the story was passed on by friends, and a map of the West based on Bridger's travels exists bearing the following note: "The original made by James Bridger for Col. Wm. Collins. First drawn in the earth or sand with a stick—then in detail on the skin of an animal with charcoal, then given to Col. W. O. Collins by Bridger; Collins then made this exact map."

In these early decades of the nineteenth century wagons and settlers were slowly starting to venture west across the Great Plains towards the distant Rockies. By the 1830s the trickle had become a migration. The settlers had maps with them but these were thoroughly inadequate: few showed even the most essential passes. So in 1838, in response to demand, the federal government organized afresh the Corps of Topographical Engineers to aid the movement westward by making better maps. From now on until the Civil War, exploration of the West lay in the hands of dashing lieutenants and captains trained at West Point.

The most celebrated was undoubtedly John Charles Frémont, "self-made cavalier," in Goetzmann's words, and one of the few not to be a West Point alumnus. Tough, handsome, brilliant, patriotic and enthusiastic to the point of foolhardiness, Frémont was something of a Hollywood hero from a classic Western. Appropriately, he would later enter politics, helping to found the Republican Party and becoming its first Presidential candidate in 1856. With a senator from Missouri, Joseph Hart Benton as his mentor, Frémont epitomized the unbreakable link between mapping and politics in the US, inaugurated by Lewis and Clark. On Christmas Day 1824 Benton had sat at the eighty-one-year-old Jefferson's feet at Monticello and talked

John Charles Frémont in 1842, planting the flag of the United States on the peak of what he thought was the highest mountain in the Rockies. Frémont included this illustration in a rushed report to Congress that had a tremendous impact on Americans thinking of moving west to settle. He subsequently led several more expeditions that produced the first accurate maps of the West, made a fortune in the California Gold Rush, and stood as the first Presidential candidate of the Republican Party.

visions of expansion in the West. Sixteen years later he picked the young Frémont, fresh from two years of disciplined experience exploring Minnesota and the Dakotas with the immigrant French scientist Joseph Nicollet, to realize his dream. Frémont was to chart the Oregon Trail, scouting passes through the Rockies and determining where to place outposts.

As his topographer Frémont chose a red-faced German named Charles Preuss, who was as prosaic and cautious a character as Frémont was romantic and ambitious. Ten years older than Frémont, Preuss had come to the US from Prussia in 1834 after studying geodesy and working as a government surveyor. While Frémont won the glory (ably assisted by his wife Jessie, Benton's daughter, who helped write his reports), Preuss produced the maps: the first accurate pictures of the West, to a standard comparable with topographical surveys in India in the 1840s, if not with the exactitude of the Great Trigonometrical Survey.

Frémont's first expedition turned out to be more of a lark than serious exploration. Instead of mapping a trail through the Wind River Mountains, he climbed what he took to be the highest peak and planted a homemade American flag on it. This he reproduced in a rushed report to Congress, along with some musings about the sublime grandeur of the scenery, revealed to him by the flight of a "weary little brown bee" he found buzzing on the heights. The gesture was shallow but its impact on the public tremendous. Within a year the Oregon Trail was alive with migrants, though still barely marked on the map.

"I fancied I could see Frémont's men, hauling the cannon up the savage battlements of the Rocky Mountains, flags in the air, Frémont at the head, waving his sword, his horse neighing wildly in the mountain wind, with unknown and unnamed empires at every hand," wrote Joaquin Miller, one of the settlers of the Oregon, recollecting his boyhood on an Ohio farm in the tranquillity of old age.

No explorer, not to say politician, could have asked for more impact from his words. Senator Benton was delighted with Frémont, though Preuss confided to his diary (in German) that he thought him childish. Next year Benton got Frémont command of a more substantial expedition—the one for which history remembers him. Frémont's orders were to link the reconnaissance of 1842 with existing American charts of the Pacific coast, so as to give a "connected survey" of the interior of the entire continent. He left St Louis in early spring 1843, with Preuss and thirty-eight men. The Oregon Trail was already busy with people seeking Frémont's vision of the West. Ahead was a trail-blazing Californian party and beyond them a caravan making for Oregon. Behind were Belgian priests bound for a mission station in the Bitterroot Mountains, an itinerant politician and future governor of Colorado, and a Scottish laird out for his last hunt along the Seedskeedee river before returning to his castle in Scotland.

The expedition turned south and crossed the Rockies into Colorado, went north into Wyoming, and then south again to the Great Salt Lake, imagined by Frémont to be unexplored. Its "still and solitary grandeur" stretched "far beyond the limits of our vision. . . . I am doubtful if the followers of Balboa felt more enthusiasm when from the heights of the Andes [sic], they saw the great Western Ocean." Frémont's word for the area around the Lake—"bucolic"—inspired Brigham Young to lead his Mormons there and settle. Next the expedition turned north towards the Oregon, following the Snake downstream to the Columbia. Frémont's intention was to enter Fort Vancouver by river but Preuss refused to make himself "presentable" by shaving his beard. According to a marginal note in his diary by Mrs Preuss, "Here Frémont became so mad that he wanted to challenge Carl to a pistol duel because the

Gold was first struck in California in 1848, attracting hordes of fortune-hunters, as well as a few scientists like the brilliant Josiah Dwight Whitney, later to become the first director of the California Geological Survey. Whitney wrote to his brother that year: "California is all the rage now. . . . We are already planning to secure the geological survey of that interesting land, where the farmers can't plough their fields by reason of the huge lumps of gold in the soil."

Right The construction of the Union-Pacific Railroad across the United States gave a tremendous impetus to the surveying of the West. Vigorous campaigning for different routes had begun in the 1840s and lasted for twenty years.

latter did not want to cut off his beard." Somehow the two men patched up their quarrel and the epic journey continued south into California, eliminating en route the hypothetical river Buenaventura, supposed to flow west from the Rockies to the Pacific.

They marched through western Nevada, along the edge of the arid Great Basin between the Sierras and the Great Salt Lake. Frémont had intuited the existence of this vast depression from his observations at the Lake. Across that space on the map Preuss later drew back in Washington DC, Frémont inscribed the legend:

> The Great Basin; diameter 11° of latitude, 10° of longitude; elevation above the sea between 4000 and 5000 feet [1220–1520 m]: surrounded by lofty mountains: contents almost unknown, but believed to be filled with rivers and lakes which have no connection with the sea, deserts and oases which have never been explored, and savage tribes which no traveller has seen or described.

Further south, in those "lofty mountains," the Sierras, winter caught them and almost brought disaster. One man had hallucinations, wandered off and never returned; the Indians in the party began singing their death songs; and Preuss got lost for three days, in which hunger reduced him to placing his hand in a nest of ants and then licking them off. But a Californian spring eventually restored even his gloomy spirits, ready for the trek back through the Southwest to St Louis.

Preuss did not join the irrepressible Frémont in the field again until 1848, but concentrated instead on making maps of the West. His third and last was entitled "Map of Oregon and Upper California," but in fact it showed all the territory west of the 105th meridian. The Californian gold-fields appear on it for the first time as "El Dorado or Gold Region," as does the name "Chrysopylae or Golden Gate" at the entrance to San Francisco Bay. California seems to have gone to Preuss' normally level head, because he returned there to join Frémont in a crazy scheme aimed at finding a route for the much-touted railroad across the southern Rockies. The expedition foundered in bitter cold and deep snow and some of the survivors were accused of cannibalism. Preuss was lucky to escape alive, his health battered. After this he could never settle down anywhere. Five years later he was back in Washington, where he hanged himself from a tree. Frémont stayed in California and made a fortune in the Gold Rush, entered politics as a senator, became a Civil War general (having earlier been court-martialled for insubordination in the Mexican War), lost his wealth through unwise railroad investments, and ended his days as territorial governor of Arizona, dying at seventy-seven, in 1890.

The information from Frémont's expeditions, along with that of many other members of the Corps of Topographical Engineers and all earlier sources, was sifted and consolidated in the 1850s into a remarkable map of the West, after which "subsequent efforts," says the historian Carl Wheat, "may properly be deemed merely filling in the detail." Its author was a young officer, Gouverneur Kemble Warren, a graduate of the 1850 class at West Point, with the ambition and good looks of a second Frémont. Completed in 1857, Warren's map was published as part of Volume II of the monumental thirteen-volume series *Pacific Railroad Reports*. It was, remarked Warren, unconsciously echoing James Rennell and his *Map of Hindoostan*, "the culmination of more than three centuries of effort to come to grips with reality in this vast and complex area."

Warren's map—and indeed the Union-Pacific Railroad, which was finally finished in 1868—paved the way for a new age in the exploration and mapping of the

West: the rise of the civilian scientist and the eclipse of army exploration. This era opened in California in 1860, with the founding of the California Survey, whose purpose was to make an accurate and complete geological survey of the state, and "to furnish ... proper maps, and diagrams thereof, and with a full and scientific description of its rocks, fossils, soils, and minerals, and of its botanical and zoological productions." No reference was made to the Survey's potential utility in exploiting all these resources. Instead, its director, Josiah Dwight Whitney, informed the state legislature that "it was not the business of a geological surveying corps to act to any considerable extent as a prospecting party." Such high-mindedness was certain in the longer run to bring many of the new state's settlers into conflict with its author.

Half Dome in Yosemite National Park, California, which fascinated Whitney when he first visited it with the California Geological Survey. Despite opposition in California, Whitney and others finally persuaded President Lincoln to set the entire area aside for conservation and enjoyment by all Americans, and in 1864 Yosemite was declared a State Park.

Whitney was an easterner of aristocratic bearing and with a reputation for wit, whose education and experience were virtually unparalleled in America. After graduation from Yale, a Phi Beta Kappa, he had studied for years in France and Germany, latterly with the great chemist Baron von Liebig. Back in the US he spent some years on surveys in New England and around the Great Lakes, producing his classic book *The Metallic Wealth of the United States*. California had interested him from the days of its first gold rush. "California is all the rage now," Whitney wrote to his brother from Boston, "and poor Lake Superior has to be shoved into the background. We are already planning to secure the geological survey of that interesting land, where the farmers can't plough their fields by reason of the large lumps of gold in the soil."

The gold itself did not fire Whitney's imagination, but its pattern of occurrence did. When he at last reached California, he quickly discovered that the maps in the *Pacific Railroad Reports* and the charts of the coast were totally inadequate to his

needs. He and his first-class assistants, including Clarence King, were forced to produce topographical maps of the state in order to get at its geology. Within twenty years their efforts led to the founding of a national geological survey (with King as its first director) that has since provided most of the maps used by every American, from the soldier and the specialist to the ordinary tourist. With the California Survey, US cartography at last began to match the standards of accuracy of national surveys in Europe and India.

For seven years Whitney and his men threw themselves into mountain-climbing in their efforts to get the measure of California. The 1864 season took them to the High Sierras, the first full-scale expedition to reach these peaks. Hauling their theodolites and barometers up to the gaunt summits, they added an area to the map of the state "as large as Massachusetts and as high as Switzerland." They also surveyed Yosemite and became fascinated by its geological formation. The grand sweep of the landscape took such a hold on their imaginations that, along with others, they urged President Lincoln to grant Yosemite to the state of California as a place set aside for enjoyment by all Americans. This was the genesis of the idea of National Parks, a natural offspring of Whitney's scientific spirit and lack of commercial drive. But it soon raised the hackles of those with "special interests" in the state legislature, who saw Whitney as an easterner who thought he knew better than they did what California stood for. He was clearly trying to "curtail the big bonanza," as Goetzmann puts it; and they were out to control him. In 1864 the legislature had the Survey restricted to "a thorough and scientific examination of the gold, silver and copper producing districts of this State." Though it continued under Whitney's leadership, he gradually lost heart and returned east in 1868.

F. V. Hayden, leader of the Hayden Survey of the Yellowstone in 1871, photographed by William Henry Jackson. Hayden through his surveys, and Jackson through his gripping photography, probably did more than anyone else to create the "tourist's West" in the nineteenth century; but their work also had serious scientific value, allowing Americans to see the real West for the first time.

The debate between exploitation and conservation continues, now at both national and international levels. So does another debate that stirred first in the American West about 1870: the question of tourism. F. V. Hayden, the surveyor of what in 1872 became the Yellowstone National Park, was the earliest surveyor to employ photography to publicize his work. More and more, the photographs he handed out to congressmen came to resemble propaganda. They were taken, in the first instance, by William Henry Jackson, an ex–Civil War combat artist who shot to fame in 1869 by photographing America from a perch on the cowcatcher of engine No. 143 of the new Union-Pacific railroad. Soon Jackson's masterly capture of the West on photographic plates and stereopticon slides had revolutionized its image in the rest of the United States.

Jackson and Hayden together probably did more than anyone else of the time to create the tourist's West in the nineteenth century, thereby laying the foundations for its steady despoliation in the present century. At the same time the new technology brought multiple benefits of utility and lasting value. Using photographs, the geologist could study exposed strata in his laboratory; the planner in St Louis or Washington DC could get an authentic idea of the barren, rocky landscapes facing the new settler; the ethnologist could see what Indians really looked like, shorn of romantic and racist description. The camera brought a new realism into America's understanding of its vanishing frontier. The year 1869 had also witnessed the disappearance of the last significant geographical mystery in the United States (excepting the new territory of Alaska). What was the course of the Colorado river? Various explorers and trappers, beginning with the Spanish, had crossed it at various points. A few had even glimpsed it as a ribbon of brown water at the bottom of the Grand Canyon. But no one had travelled and mapped its length.

John Wesley Powell, the intrepid one-armed pioneer of the Colorado, combined scientific method with missionary fervor and political acumen. He reminds one a little of Sir George Everest. The son of a frontier Methodist preacher, Powell had made long trips down the Ohio and the Mississippi and its tributaries during his twenties, where his lifelong interest in natural history developed. During the Civil War he had served the Union, lost an arm at Shiloh, gained a friend and supporter in General Grant, and learned how to handle government bureaucracy. In 1867 he led his first expedition to the Rockies, a collecting trip for the museum he had founded back home in Illinois. The enterprise was funded by a diverse alliance of western academics and railroad companies, supplemented by Powell's own money. On Pike's Peak, near Colorado Springs, Powell conceived his great plan.

The boats for the Colorado expedition were built in Chicago out of oak to Powell's specification, with watertight compartments at either end to maintain buoyancy in even the roughest water. On 24 May 1869 he and his party set off down the upper Colorado, known as the Green River. Nothing was heard of them for thirty-seven days, long enough for a confidence man to make a brief career as the sole survivor of the supposed wreck. During that time whole days were spent in shooting the worst stretches of rapids, as they passed through Flaming Gorge, Horseshoe Canyon, Kingfisher Canyon and Red Canyon, where they came across the message "Ashley 1825," painted on a rock in midstream: presumably the handiwork of a long-forgotten mountain man.

Before entering Lodore Canyon, Powell was able to climb out and gaze down nearly half a mile to the river below. "I can do this now," he wrote, "but it has taken several years of mountain climbing to cool my nerves, so that I can sit with my feet over the edge, and calmly look down a precipice 2000 feet. And yet I cannot look on and see another do the same. I must either bid him come away, or turn my head." About a month later Powell found himself on a steep cliff above the river frantically clawing whatever he could catch onto with the fingers of his only hand. He had climbed up there and become trapped. Above him Bradley coolly removed his long johns and lowered them to Powell in the nick of time. Gripping them by one hand, Powell was hauled to safety.

On 10 August, after passing the point at which a Spanish explorer had crossed the Colorado in 1776, the boats reached the junction with the Little Colorado, the entrance to the Grand Canyon. They were three quarters of a mile (1.2 km) inside the Earth, Powell noted. "We are now ready to start on our way down the Great Unknown. ... We have an unknown distance yet to run; an unknown river yet to explore. What falls there are, we know not; what rocks beset the channel, we know not; what walls rise over the river, we know not. Ah well! we may conjecture many things. The men talk as cheerfully as ever; jests are bandied about freely this morning; but to me the cheer is somber and the jests ghastly."

Into the foaming current they plunged, racing for the next week or two from one bone-jarring crash to another, or spending hours edging the boats past impossible rapids from narrow shelves of rock. The last stretch proved to be the worst: they were faced with a series of rapids, beginning with a tremendous waterfall 18–20 ft (5.5–6 m) high. Two of Powell's party now elected to take their chances on foot. The rest agreed to trust Powell and the river. The daring Bradley went first. As he and a boat vanished in the torrent, Powell thought he was gone for good. But he emerged quite quickly, standing on his half-submerged boat and waving his hat. Three days later they came upon some Mormons fishing in the Colorado and keeping an eye out for debris from the expedition they had been warned to expect. Above them, on the

The first expedition down the Colorado river, by John Wesley Powell and party, 1869. Braving Indians, fierce rapids and an uncertain route, Powell followed the Colorado from its source in the High Rockies through the Grand Canyon to the Gulf of California. Scientifically he gathered little, but his second expedition produced a survey of southern Utah, northern Arizona and north-western New Mexico that remained the basis of maps for nearly a hundred years.

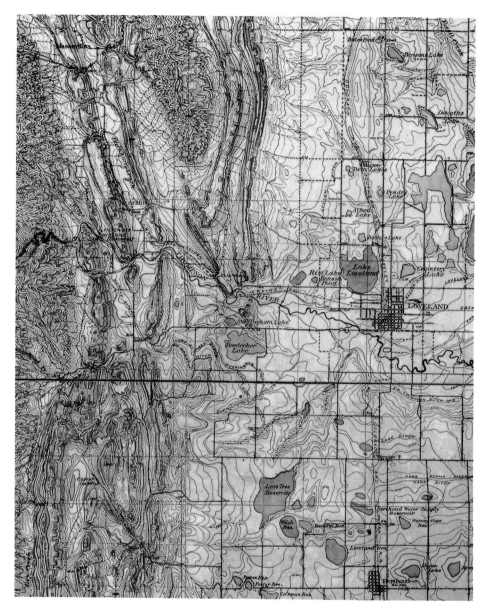

Part of Colorado surveyed by the US Geological Survey in 1905–06. John Wesley Powell, director of the USGS until 1894, laid down the principles of a complete survey of the US in rectangular sections. He thought that the work could be finished in twenty-four years; in fact his dream was not fulfilled until the 1980s, a century later, with the help of lasers, satellites, and other late-twentieth-century technology.

Shivwits Plateau, unknown to Powell, his two overland colleagues lay dead, felled by Indian arrows.

Scientifically, this first expedition was a virtual failure; even its astronomical data were lost with the murder of Powell's men. But Powell was a hero. With official support, he determined to make a second, longer expedition, properly equipped to survey the river and the surrounding "Plateau Provinces" left blank on Warren's map: southern Utah, northern Arizona, and northwestern New Mexico. Instead of measuring latitude by sextant, they would use a zenith telescope, and instead of the method of lunar distances for longitude, telegraphic time signals from an astronomical station at Salt Lake City. A proper baseline would be laid out, triangles surveyed through the clear desert air, and survey mounds constructed.

Some of the most remote areas of this survey, which was conducted mainly by Powell's brother-in-law A. H. Thompson, remained the basis for maps of the region nearly a hundred years later. In the course of its work the last unmapped river and mountain range in the United States, bar Alaska, were finally located. They were named after the 1776 Spanish explorer Escalante, the first white man known to have crossed this wilderness, and Henry, the secretary of the Smithsonian Institution in Washington DC, a friend and backer of Powell.

In writing up his results Powell permitted himself a bold overview of the geology of the region. Like Whitney in California, Powell was not concerned to collect or prospect so much as to analyse and reform. He devised the novel concepts of base level of erosion in any region—towards which its streams and rivers all trend—and antecedent and consequent valley, formed respectively before and after a period of uplift and folding of rocks. Referring to the relationship of the Colorado to the Uinta Mountains Powell wrote: "Again, the question returns to us, why did not the stream turn round this great obstruction, rather than pass through it? The answer is that the river had the right of way; in other words it was running ere the mountains were formed; not before the rocks of which the mountains are composed, were deposited, but before the formations were folded, so as to make a mountain range."

Speculations of this sort led Powell to his recommendations for reform of American policy towards the West, embodied within a few years in his *Report on the Lands of the Arid Regions of the United States*. Although this was largely ignored in its time, for the same commercial reasons that undermined the wide-ranging California Survey, it returned to haunt America in the dustbowl years of the 1930s and today forms part of everyday government cartography and thinking about land use.

Powell's crucial insight was to see that, since much of the West—two fifths of the US in fact—was arid, a scientific approach to its exploitation was a necessity, with mapping as its bedrock. Once the country had been mapped it could be classified as mineral lands, coal lands, pasturage lands, timber lands, and irrigable lands. Only then could it rightly be managed and productively developed. "Like the Puritans who came to the New World to organize their model society, the City on the Hill, Powell went into the dreaded canyons of the Colorado and among the spires and dry landscapes of the Plateau Province and emerged with his own vision of the perfect society. The age was different but the aim and impulse were the same," William H. Goetzmann concludes of John Wesley Powell.

By the time he retired as director of the US Geological Survey in 1894, about a fifth of the United States had been mapped to the standard Powell deemed essential: a series of quadrangles bounded by parallels and meridians, on various scales—from four miles to the inch (2.5 km/cm) for desert regions, to one mile to the inch (0.6 km/cm) for densely populated areas—with overlays for surface geology, land use and other scientific data. Powell had earlier told Congress that the entire continent could be covered by quadrangles in twenty-four years: a pardonable underestimate of so vast and complex a task. Not until the 1980s, a century after Powell's prediction, and with the aid of technologies of which the pioneers could scarcely dream, was the United States of America truly mapped.

10

PICTURES
OF THE
PLANET

In 1980 the veteran American mountaineer and cartographer Bradford Washburn decided to carry out a youthful dream: to map Mount Everest and its surrounding peaks as they had never been mapped before—from on high. Nearly half a century before, they had been photographed by British aviators, who had barely cleared the summit; but the only maps available remained those made from the ground by the rugged men of the Survey of India, using theodolites. Since then there had been a remarkable improvement in cameras, lenses, film and aircraft, and computers and satellites had come into existence. A definitive map of this remote region had become technically feasible. The barrier to success was political: would Nepal and China, whose border crosses the peak of Everest, permit the mission? The King of Nepal said he would, provided the Chinese agreed. Washburn visited China, and became the first foreigner to receive the right to fly over Qomolangma, as the Chinese call Mount Everest, in order to photograph it precisely.

The flight took place in a Learjet at 40,000 ft (12,190 m) on 20 December 1984. Sadly, Washburn was not on board but back in the US with his wife, who had been taken desperately ill in Nepal. In three and a half hours the rest of the team took 160 photographs that would form the basis for the next three or four years of mapmaking. These alone were insufficient, however. Two other sources of information were needed: observations from space, and observations from the ground—the latest cartographic technology, to be combined with the maps of the pioneers. The space observations, ironically, provided what is known as ground control: a network of points on the ground that can be clearly identified in the photographs, with known position and elevation. Washburn likens their importance to that of a high-quality steel framework in a big building. They came from NASA's space-shuttle *Columbia*, which had flown over Everest at an altitude of 156 miles (250 km) on 2 December 1983, with a high-resolution West German mapping camera on board recording infrared pictures that were, according to Washburn, "crystal clear."

At Zurich's Technical Institute, in the department of research photogrammetry, the images from *Columbia* were compared by computer with the British, Austrian and Chinese topographic maps of Everest made between 1921 and 1975. The computer identified map errors and permitted the selection of a hundred new

Left Mount Everest *viewed from a NASA space-shuttle at an altitude of 156 miles (250 km): a superb example of the high-resolution photography that has revolutionized the process of mapmaking. At the top can be seen a few snowcapped mountains on the Tibetan plateau; Mount Everest is at center right, above the forested foothills of the Himalayas and the low-lying Ganges plain.*

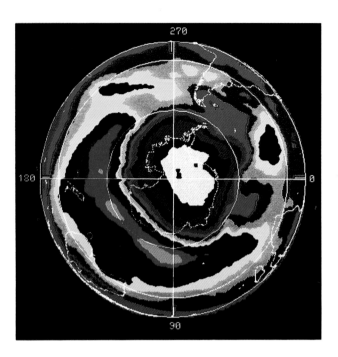

*The ozone hole over
Antarctica, measured in
October 1987 by NASA's
Total Ozone Mapping
Spectrometer (TOMS) on
board its Nimbus 7
satellite. The hole is
visible as the deep blue,
purple, black and pink
areas covering Antarctica
(outlined in white) and
beyond, which indicate
low ozone concentration.
The hole is probably the
result of pollution by
CFCs, chlorofluorocarbons.*

ground-control points, which could then be used to anchor the 160 photographs taken from the Learjet, and create a contour map. The process required two years of painstaking laboratory work. A team of gifted cartographers at the Swiss Federal Office of Topography in Berne, experienced in depicting the Alps, then had to inscribe onto the Everest map the intricate details of mountains and glaciers in the region and add the airbrushed relief shadows that make the final map look three-dimensional. Their patience, says Washburn, was "scarcely believable": often they completed only a single square inch (6.5 square cm) of map in a day.

Thousand of hours of labor later, the printing plates were ready to go to the US headquarters of the National Geographic Society, which had supported the project from the start. Here a final exacting stage was necessary: accurate naming of the peaks and glaciers—a controversial subject ever since the mid-nineteenth-century argument over possible Nepali and Tibetan names to replace the English form, Mount Everest. Careful study of all known place-names, in collaboration with experts from Nepal and China, led to the use of as many as four languages to name some features. The whole operation would surely have boggled the mind of Sir George Everest, even if its astonishing end result is something he could have comprehended and appreciated. Today's satellite photography and Everest's ray-tracing in the 1830s bear as much resemblance to each other as an automobile does to a bullock-cart. Everest would have understood the teamwork though. Nine nations participated in the project: the United States, Switzerland, Nepal, China, Britain, West Germany, Sweden, Yugoslavia, and Japan. "Truly this was a project of people," wrote Washburn, "each an expert in his own field, working in concert and respect."

Washburn's map of Everest is, in a sense, both the culmination of centuries of cartography, and also a harbinger of future possibilities in mapmaking. It is a superbly detailed and accurate rendering of an area of the Earth's surface, in the tradition of the *Carte de Cassini* and the Survey of India. But the computers that made it possible—as distinct from the skill and artistry of the Swiss who drew it on paper—represent a total break with that tradition. Without at all diminishing the utility (and beauty) of paper maps for a range of purposes, computers have taken mapping into a new electronic era, in which a map is an extremely complex array of data stored in a "black box," to be manipulated at will. From now on many maps will be "invisible," says John Garver Jr., Chief Cartographer of the National Geographic Society. They will be kept on disc, called up, and wiped from the screen once used.

After computers, information-gathering from satellites, used to compile the Everest map, is the second development that has radically changed the nature of mapmaking. The idea of aerial photography was nothing new, but the fact that satellites see the planet, not just a small area of it, the great variety of ways in which they can look at the Earth, and the capacity of computers to process the staggering amounts of data satellites collect, together mean that cartography is no longer national or continental in scope, but truly global. In the late twentieth century state-of-the-art mapping can visualize almost every aspect of our physical world. Examples include Mount Everest, the floor of the Pacific Ocean, the shape of the ice-cover and the continent below the ice in Antarctica, plankton concentrations in the Gulf Stream, the hole in the ozone over Antarctica, the probability of earthquakes in different parts of California, the burning of forests in the Amazon Basin, the flight pattern of airliners over the US at any moment, the incidence of AIDS in a New York City suburb, or the likely impact of global warming on the coastline of Europe. In addition, these maps can be altered instantaneously, either to take account of new information, or because the operator wants to use the map for a different purpose. Computer-based maps offer the endless scope for electronic overlays that John Wesley Powell first imagined and introduced on a small scale in his paper maps for the US Geological Survey a century ago.

The key lies in the application of what are called Geographic Information Systems—GIS—an idea made practical by the immense storage and analytical capacity of computers. A USGS cartographer explains: "A GIS gives you much more capability in analyzing map information because you're not constrained to the information that's on a paper map. We can collect information in real time and have that displayed on a computer screen as it's being collected." He gives an example: a USGS analysis of an earthquake-prone area near Salt Lake City, Utah. Geological information about the area was fed into a GIS to create a map showing the stability of the land surface during an earthquake. The emergency response system for Salt Lake City was then overlaid on this to see how it would cope in an earthquake. "We saw some very dramatic examples of areas that would be difficult to serve with emergency services."

Such GIS analyses need not be static. The computer can combine two sets of data, such as the images taken from above Everest and suitably digitized topographic information from old maps in the map we have just discussed, to produce a perspective view. When a series of such views are played sequentially, they give the appearance of a fly-around. US geologists are already "flying" over terrain, examining features and especially the texture of hills where landslides occur, in the hope of predicting or even preventing them.

Most of the information fed into a GIS, and all of the information produced by satellites, depends on what is known as remote sensing. Refinements in remote sensing of the Earth have been responsible for the major advances in twentieth-century cartography. Aerial photography, sonar, radar, seismic measurements, and satellite photography, all involve remote sensors; so does the human eye. Different remote sensors detect different portions of the spectrum of electromagnetic radiation. The eye is sensitive to visible light: radar to microwaves: satellites to microwaves, infrared, visible and ultraviolet radiation, or a combination of all four, depending on the scanners the particular satellite carries. These scanners may be passive or active: in other words they may measure radiation falling upon them from other sources, or they may measure radiation emitted by themselves and reflected back to the scanner by material on or below the Earth's surface.

All material, whether solid, liquid or gaseous, has a characteristic signature: it absorbs incident radiation at certain wavelengths while reflecting the radiation at other wavelengths. The phenomenon of color is a simple example: the sea appears blue because it reflects the sun's light strongly at the blue wavelengths of the visible portion of the electromagnetic spectrum. At other wavelengths it may strongly absorb incident radiation in ways that depend on its surface temperature, roughness and plankton concentration. Thus two objects may "look" alike in one wavelength, yet be quite distinct in another. Scanners must be capable of measuring these differences in reflectance of radiation with as much precision as possible. It is they that enable a satellite image to distinguish, say, ice from water around Antarctica, sandstone from limestone in the Yellowstone Park, or plankton-rich water from plankton-poor water in the Atlantic.

The first satellite was the Soviet Union's *Sputnik I*, launched in 1957. The first communications satellite was *Echo I*, launched by the US in 1960. Neither satellite

Europe and North Africa photographed by the European Space Agency's NOAA-9 weather satellite. A severe storm covers most of the British Isles and cloud can be seen over Norway, southern Italy, south-eastern Europe and North Africa. Such storms have been detected during their formation ever since weather-satellite information became routine in 1966.

carried scanners. One of the largest artificial bodies ever orbited, the size of a ten-story building, *Echo* reflected radio signals from coast to coast off a thin coating of aluminum on a great balloon of Mylar with only half the thickness of cellophane. *Telstar*, which followed in 1962, carried transponders capable of amplifying radio signals thousands of times before relaying them, but had the disadvantage that it could do so only when its orbit brought it into line between sender and receiver. This drawback was resolved by *Early Bird*, launched by the US in 1965, which could be parked in geostationary orbit,—that is, at a height and speed that kept it fixed with respect to the Earth. Satellite telephone and television now became a reality.

The satellite with eyes that made global mapping feasible was launched by NASA in 1972. *Landsat I* orbited 570 miles (920 km) above the Earth, fourteen times a day, peering down on a 115-mile-wide (185 km) band of the Earth's surface. A television camera on board could distinguish objects about the size of a football field. The multispectral scanner operated in the visible and infrared spectrum. Each scan line contained 3240 discrete pixels (picture elements), equivalent to a ground resolution cell of 184 × 259 ft (56 × 79 m). A single Landsat scene (115 × 115 miles; 185 × 185 km) consisted of 2340 scan lines. Each scene was made up of more than thirty million individual observations. Since the scanner took about 25 seconds to complete a scene, the rate of data generation and accumulation was staggering.

Satellites, and Landsat and its sucessors in the 1970s and 1980s in particular, opened up the possibility of measuring, analyzing and beginning to understand global processes. For the first time, cartographers could do synoptic or specific surveys at will, and were able to repeat measurements over long periods, in order to monitor the behavior of the planet. To start with, the new techniques perhaps seemed a marvellous luxury; soon, however, they had become a lifeline, vital to our long-term survival, as we attempt to control our own dislocation of global systems.

One of the earliest applications of satellites lay in navigation. US scientists listening to the beeps of *Sputnik I* found they could predict from them the precise position of the satellite, and thus their own location on Earth. An expert navigator, equipped with a sextant, can get his latitude with an accuracy of one mile (1.6 km) at dawn or dusk, assuming a clear sky; a satellite navigator can tell where he is to within 100 yards (90 m), night and day, rain or shine. "This is the most striking innovation in navigation since the compass for giving orientation, the sextant for determining latitude, and John Harrison's chronometer for determining longitude," according to Robert Dietz of the US National Oceanic and Atmospheric Administration. Its only failing, like all sophisticated technology, is that it might break down, stranding navigators who are unable to use a sextant.

Weather forecasting and climatic maps were another, now familiar, output of satellites, in this case called TIROS and GOES, Geostationary Operational Environmental Satellites. One of the great advantages of these satellites for meteorologists is their ability to observe conditions over areas where other information is lacking, especially the oceans. Since 1966, when weather-satellite information became routine, no tropical storm has formed undetected anywhere in the world. For fishermen they provide data on sea temperature of immense value, since the fishermen know the temperatures fish prefer but not where those waters can be found. With the advent of satellite temperature charts a skipper can observe, say, a warm eddy moving into the Hudson Canyon off New York City—perfect for mackerel, squid and scup—hurry there and make a big catch.

Landsat pictures have the widest range of immediate uses: in agriculture, soil science and forestry, in geology, geophysics and mining, and in hydrology and

Right *An American grizzly – one of many unexpected beneficiaries of modern mapping techniques. Satellite imagery sensitive to different kinds of vegetation has helped zoologists identify the habitat of grizzlies and predict where they are likely to roam. It turns out that their favourite foraging ground lies at about 8800 feet (2680 m).*

Right *Fires in the Yellowstone National Park, 1988. Satellite overviews of these huge fires, fed by nearly ninety years of unburnt kindling, and of the state of the Park after the conflagration, have helped shape future official policy towards fires. The size and location of burn patches can be monitored in relation to new growth in order to understand more fully how the Park regenerates itself.*

oceanography. Guided by Landsat imagery analysts have discovered lakes unknown on any maps, a new islet off Canada's Atlantic coast (named Landsat Island) and an uncharted reef in the Indian Ocean; located hundreds of geological features such as great crustal fractures; found oil in Sweden, tin in Brazil, and uranium in Australia; and mapped routes for railways, pipelines, and electric power lines.

Two applications of Landsat in America's Rocky Mountains show how it has changed environmental science. In Idaho, Wyoming and Montana wander probably fewer than a thousand grizzly bears, the remnants of tens of thousands of the bears that once roamed the American West. In the early 1970s the grizzly's biology was quite well understood, but not its habitat, the kind of land and elevation it preferred. "By understanding the grizzly's habitat in a specific area of wilderness, could we not more accurately estimate its population?" asked the zoologist John Craighead. And then go on to make better-informed decisions about grizzly bear management? The problem was how to relate the variety of terrain in an area of wilderness where grizzlies were known to be active to the ten million mountainous acres (40,500 sq km) where they might wander.

Landsat pictures, computer-processed to distinguish a range of terrains such as vegetated rock, subalpine parkland, alpine meadow, sandstone or limestone, proved the solution. Two summers' fieldwork in the chosen wilderness provided a "ground-truth" check of the Landsat imagery: Craighead could now match an area on an image with a particular type of terrain. He could then predict where in Idaho, Wyoming and Montana a grizzly was likely to be found, and estimate how much of such habitat existed (and hence how many bears might live in it). Sightings of grizzlies, and secondary evidence such as tracks, feces, and diggings, confirmed that the grizzly's favourite foraging ground is a vegetated rock complex at about 8800 ft (2680 m).

In the Yellowstone National Park, in Wyoming, the matching of Landsat imagery and ground-truth observation has helped to explain how the Park regenerates itself after fires, and to suggest how these might be better managed than in 1988, when they burned nearly three quarters of a million acres (3040 sq km) inside the Park: the result of the accumulation of ninety years' worth of kindling during the period in which the Park authorities believed in putting out natural fires rather than letting them burn. The policy since the 1970s has been that fires are neither bad nor good, provided they do not threaten lives or property. "An unburned forest is good for some species and a burned forest is good for other species," says Don Despain, research biologist and expert in fire ecology. "Some of the grasses don't flower unless the forest burns. Then they profusely seed and some of the seed goes to the burned area.... Nutrients released from the ashes go into the soil and the plant response following the fire is much more luxuriant. The larger animals—elk, deer, and moose—perfer to eat the regrowth. It's nutritionally better for them, and I guess it tastes better to them as well." By examining satellite images of the 1988 fires and their aftermath Despain hopes to learn how the variety and size of burn patches influence forest regeneration, and why exactly so many acres burned. The conclusions he reaches may also help the commercial foresters outside the Park to manipulate the environment to produce the forest products they want.

Needs and expectations for satellite imagery are often cruder elsewhere. In the Third World, Landsat has been an invaluable help to governments lacking basic maps of impassable terrain. In Bolivia, for example, more than four hundred bush pilots a year were crashing while looking for incorrectly mapped villages and running out of fuel, or colliding with mountains at night that were not supposed to be there.

Landsat, besides reducing the crashes by allowing correction of the maps, aided civil-engineering projects and mineral prospecting. It helped to identify potentially lode-bearing rock outcrops and streams that might permit canoes to reach them. Sought-after tin frequently occurs in ancient streambeds and granite outcrops, which are not visible even to a person standing on them. Landsat images, suitably processed to enhance the false colors, can discriminate between normal granite and tin-bearing granite because, it appears, the tin beneath alters the vegetation on top.

Brazil, too, has taken advantage of Landsat, but it was a different technique, side-looking airborne radar (SLAR), that permitted the first mapping of the Amazon Basin, only two decades ago. Originally developed for military use, SLAR improved on the revolving antenna of earlier radar systems by attaching a larger, static antenna to much of the length of an aircraft, and pointing its beam sideways. Since a radar system's resolving power increases with the size of its antenna, as does a telescope's with the size of its lens, SLAR could see more detail, and over a wider area, than earlier systems. It proved to be ideal for flying long, overlapping swathes over the unmapped rain forest. Unlike aerial photography, it was unaffected by the constant cloud cover, and so work could continue day and night. The images produced were not only less distorted than aerial photographs, but more sharply etched for relief. "An SLAR picture has no sense of color, but a very strong sense of form," says Homer Jensen, the US pilot whose company was contracted by the Brazilians to tackle the Amazonian void. "It's like the shadows and apparent darkness you get from viewing in the evening light."

One of the discoveries was the hilliness beneath the forest canopy. Lowlands extend over only about twenty per cent of the Basin, mostly along the Amazon and its tributaries. The survey exploded the myth that the Amazon was basically flat;

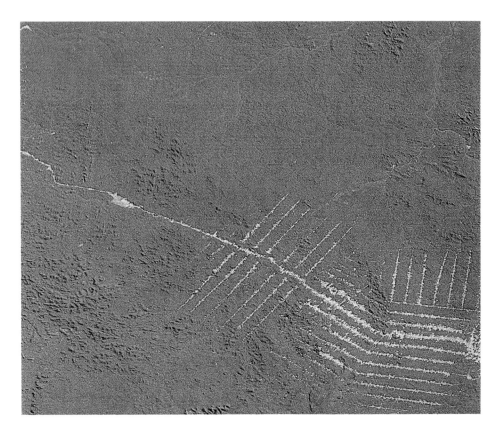

New highways slashed out of the rainforest of the Amazon Basin, photographed by Landsat (left) *and on the ground* (right). *The highways were built in preparation for settlement in western Brazil of "slash and burn" farmers, relocated by the government from other impoverished regions of the country. Satellite imagery has confirmed the alarming rate of deforestation in the Amazon.*

explorers such as La Condamine in 1744 had obviously generalized from their experience on the rivers. The positions of unexplored tributaries were also often very inaccurate on the existing maps, being mainly guesswork. SLAR detected the human errors and charted the streams correctly for the first time. It also found an unknown river more than 100 miles (160 km) long, located mountain ranges far from their supposed positions, and discovered that a region marked as a national forest reserve was a savannah. Today, as the forests he helped to survey are incinerated by cattle ranchers and settlers, Jensen admits to some doubts about the project. "It was a phenomenal success technically—we completed the maps and they're useful. The impact was not brought about by those maps, but it was facilitated by them—made easier and quicker. Hopefully, the fact that you have complete information permits intelligent people to make decisions which will protect, as well as aggressive people to make decisions which will destroy."

At the time of Jensen's surveys the Amazon Basin was the planet's last uncharted land frontier. Below the oceans whole worlds have been opened up to science in the years following the Second World War, mainly as a result of remote-sensing technology improved in that war. In 1950, knowledge of the sea-bed and the mechanism of ocean currents had not greatly advanced on what was known in the late nineteenth century, which was encapsulated in the fifty volumes of reports published between 1890 and 1895 arising from the remarkable expedition of the British ship *Challenger*. Its voyage, in 1872–76, covering more than 78,000 miles (125,530 km), set a standard for all later oceanography. *Challenger* made 364 stops to collect marine life and sedimentary samples, and to take seafloor soundings with lead and line. One of these, in the Pacific, reached a depth of about 33,000 ft (10,060 m), a

Atlantis II, *a scientific research ship belonging to the USA's Woods Hole Oceanographic Institution. Its predecessor* Atlantis *was responsible in 1947 for discovering the Mid-Atlantic Ridge, a vast underwater mountain range whose existence was the first hard evidence that the controversial theory of continental drift was true.*

record at that time. Facts collected then were still in use in the 1980s—among them, water-temperature readings that suggested the existence of a high mountain barrier down the middle of the Atlantic Ocean.

In 1947 the *Atlantis*, using the most powerful depth-sounder then available, proved the existence of such a mountain range beneath the sea. Led by scientists from the Woods Hole Oceanographic Institution in Massachusetts, with support from the National Geographic Society, the expedition sailed east from Bermuda towards the Azores. At first, the depth-sounder showed irregular terrain, then an abyssal plain that was absolutely flat, a vast mud floor 17,400 ft (5300 m) below the surface. Then bumpy foothills began to appear, followed, in mid-ocean, by a "wild and jagged realm of mountains, rank on rank of them, broken by huge valleys and canyons," in the words of *National Geographic*. The peaks were up to 10,000 ft (3050 m) high, a mile (1.6 km) below the surface of the Atlantic. Zigzagging back and forth across the mountains, *Atlantis* took dredge samples which revealed that the rocks seemed to be volcanic in origin.

For Bruce Heezen, a young geologist on board, the discovery of the Mid-Atlantic Ridge began a lifelong career as a pioneer mapper of the ocean floors. He and Marie Tharp, his drafting assistant, began to collect all the depth recordings they could obtain, from all the ship tracks and oceanographic centers of the world. This data enabled them to construct sea-bottom profiles—among the first such profiles to be drawn across the Atlantic—and eventually sketch maps that showed the ocean floor in three dimensions. The first volume of these was published in 1959 under the title *The Floors of the Oceans: I. The North Atlantic*. The maps were later redrawn by an Austrian artist of Alpine scenery, Heinrich Berann.

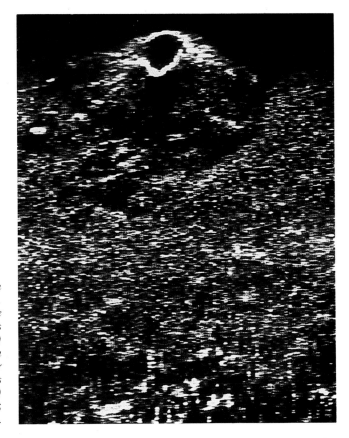

A submarine volcano on the East Pacific Rise, observed by long-range side-scan sonar. It lies nearly 5000 feet (1500 m) high on the Rise, at a water depth of over 13,000 feet (4000 m). Its crater is 1.25 miles (2 km) across and its base is 6.25 miles (10 km) across.

Meanwhile, in 1952, Marie Tharp had made a discovery of her own that would shortly become historic. "It took me eight months to convince Bruce it was really there," she says today. "He didn't publish it until 1956." She had suggested to Heezen that the deep, V-shaped valley she was seeing on each profile of the Mid-Atlantic Ridge might run along the entire center of the Ridge. Heezen was sceptical, until he took the data for earthquakes in the Atlantic that colleagues were then studying, and plotted the occurrence of the epicenters on the charts Tharp was drawing. Suddenly, he saw the light: the earthquakes were taking place in the rift valley of the Ridge. A parallel investigation by Heezen of breaks in transatlantic cables coincided with earthquake data: the cables broke over the rift valley.

Between 1956 and 1960 US and British oceanographic expeditions trailed the world's oceans with depth-recorders, tracing their ridge systems. These were found to run along the center of the Indian Ocean, as well as that of the Atlantic, linking together south of Africa and connecting with a ridge midway between Australia and the Antarctic that links to a ridge running northwards through the eastern Pacific. A rift valley was not always found at the center of a ridge; often whole segments of a ridge were offset from each other by as much as several hundred miles, along tremendous fractures in the Earth's surface, sites of oceanic earthquakes. These discoveries showed unambiguously that the surface of the Earth had once been and very probably still was in large-scale motion. The fact that the Mid-Atlantic Ridge and the ridge in the Indian Ocean paralleled the corresponding continental slopes supported this belief.

The theory of continental drift had had a rough ride through the twentieth century until this time. First advanced in 1912 by Alfred Wegener, a German

astronomer, geophysicist, and meteorologist, it had seemed to undermine both common sense and the reputations of geologists. On the basis of the impressive congruency of the two Atlantic coasts, Wegener had seriously proposed that Earth's continents had once been part of a completed jigsaw puzzle, with South America fitting snugly into Africa. He called his proto-continent Pangea, "all lands." What is more, Wegener wrote in his *Origins of Continents and Oceans*, the continents were still moving: after millions of years, the Rockies would meet Japan.

"Utter, damned rot!" scoffed a president of the American Philosophical Society in the 1920s. "You have shaken the foundations of geology," Bruce Heezen was told after describing the Mid-Atlantic Ridge to Princeton University scientists in 1957. A few years later the evidence became incontrovertible. In 1963 British ocean-ographers Frederick Vine and Drummond Matthews suggested how continental drift, or rather seafloor spreading, could account for the remarkable "zebra" maps then being made of the strength of magnetism in rocks taken from fracture zones in the Pacific floor. Instead of being random the magnetism was found to alternate in simple patterns of narrow strips, a black strip representing high magnetism, a white one representing low magnetism. Vine and Matthews knew that the Earth's magnetic field had reversed direction frequently in its history—the north magnetic pole becoming south and vice versa; the geological record said this had happened at least three or four times every million years during the past 70 million years. They therefore proposed that, at an ocean fracture, volcanic rock became magnetized in the prevailing direction of the Earth's magnetic field. After the rock cooled it retained that direction of magnetism, when the Earth's field later reversed. New volcanic rock, forcing apart the cooled volcanic rock, then became magnetized in the reverse direction. It too cooled and retained its direction when the Earth's field once more reversed. The process was then repeated, over and over. The bands of high and low magnetism at a fracture were thus a magnetic "fossil record" of the changes of direction in the Earth's field, over the period that new rock was being extruded onto the ocean floor.

During the 1960s, Wegener's continental drift theory mutated, via the evidence for ocean-floor spreading, into the theory of plate tectonics used today to explain earthquakes and the movements of the Earth's surface. The crust is said to be broken into seven major plates and many minor ones, which float on more fluid rock below. Their average thickness is about 60 miles (100 km). A younger generation of geologists easily imagines surface features in these terms. "We are taking California apart piece by piece," says Clark Blake of the USGS. "We are finding it's made up of many packets of far-travelled real estate. San Francisco is built on three different and distinct rock units that have come rafting in from somewhere in a proto-Pacific Ocean."

Speculation about the Earth's crust has a special edge of urgency in California, which sits on the San Andreas fault, the world's most famous and respected fracture zone. Not surprisingly, it was a scientist at California's Institute of Technology, Charles F. Richter, who invented the Richter scale used to measure earthquakes. That was in 1935, thirty years after the 1906 earthquake levelled San Francisco with an estimated shock of 8.3 on the scale. The latest major California earthquake, in October 1989, at 6.9 was a lesser but salutary reminder of how vulnerable California is. "The question is not whether a big earthquake is coming. The question is when," the Director of the USGS announced at the time.

The first modern understanding of earthquakes came out of the 1906 catastrophe. Henry F. Reid of Johns Hopkins University in Baltimore proposed that

A "zebra" map of magnetic anomalies at a fracture zone of the ocean floor. The pattern is generated by historical alternations in the direction of the earth's magnetic field, which are recorded like fossils in the rocks of areas showing long-term volcanic activity. Each time new rock cools it preserves the direction of the magnetic field.

Left *A map of the Indian and Pacific Oceans shows the mountains and trenches of the deep as detected by satellite measurements. The satellite records the distance between its own position and the ocean surface, which varies from place to place by up to 600 ft (185 m) because of gravity, following ocean-floor topography.*

sudden slippage and elastic rebound of crustal blocks at a fault was the mechanism: the crust bends under strain, snaps, and then straightens out. Plate tectonics theory added to this the idea that seismic disturbance occurs only close to the surface where the rocks of adjacent plates are brittle and rub against each other with friction: below about 6–7 miles (10–11 km) there is no disturbance because the rocks are hot enough to be ductile.

Seismic activity in California is being constantly monitored and mapped. Seismometers register many thousands of small earthquakes every year, and computers instantly calculate the location, depth and magnitude of an earthquake. Laser distance-ranging networks can detect changes of length, indicating change in crustal stress, accurate to about half an inch in 20 miles (1.3 cm in 32 km). Satellite measurements of crustal blocks are improving and may in time be precise enough to allow earthquake prediction. An increase in activity is unfortunately not a reliable guide to the imminence of a major earthquake in California or anywhere else. It worked in the 1975 Haicheng earthquake in China, when swarms of small shocks became vigorous three days before the main shock and massive evacuation could be ordered. But not in the Tangshan earthquake some eighteen months later, when there was no such warning and 250,000 people were officially estimated to have died.

Longer-term prediction, though less useful, rests on firmer foundations. One of these is paleoseismology: the attempt to read the future in the seismic record of the

Plate tectonics in the Red Sea and Gulf of Aden, viewed by satellite. The crust of East Africa and Saudi Arabia, floating on two different plates, has split to form the Red Sea and the Gulf of Aden. A similar but less clear-cut "fit" between the South American coast and that of West Africa led Alfred Wegener to propose his theory of continental drift in 1912.

immediate past. Kerry E. Sieh of the California Institute of Technology has identified and analyzed a history of at least twelve large earthquakes in the past fourteen hundred years, by examining a complex pattern of fractured and refractured layers of peat, silt and sand in the San Andreas fault, 35 miles (56 km) north of Los Angeles. According to Sieh, the average recurrence interval in that area is 140–150 years. The last major earthquake there was in 1857. People in Los Angeles are taking evidence like Sieh's very seriously. But everyone is aware that even if earthquake prediction became accurate to within a matter of days or even hours—and false warnings less likely—a constructive response is problematic. "We have to realize that in a city like Los Angeles the idea of evacuation is absurd," says Clarence Allen, a geophysicist at the California Institute of Technology. "Even if you could get all the people out of their lairs, God knows where they would go. So we have to be able to live with an earthquake, staying within the city."

Artificially created earthquakes have been put to good use in geology for some years, to gain information about the structure of the Earth within and below the crustal plates. The technique began in the 1920s, but the advent of computer processing in the 1950s increasingly made possible remote sensing of faint echoes of seismic waves sent deep into the Earth by explosions set off on the surface. Knowledge is nevertheless at a rudimentary stage compared with, say, our knowledge of ocean-floor geology. An important advance came from the Consortium for Continental Reflection Profiling (COCORP) under the leadership of Jack Oliver of Cornell University in the US. In 1974 COCORP began using the Vibroseis technique, in which huge 20-ton trucks thump the ground to produce vibrations of 8–40 cycles per second. Magma chambers—bodies of molten rock deep beneath volcanic areas—have been directly imaged; major faults along mountain ranges have been traced to at least 20 miles (32 km) into the crust, and the first detailed, high-resolution studies of the Moho discontinuity—the boundary between the crust and the mantle—have been obtained.

More recently, investigations have moved out to sea, using a seismic technique employed by petroleum geologists searching for deposits of hydrocarbons. British Institutions' Reflection Profiling Syndicate (BIRPS) was formed at Cambridge University under Drummond Matthews, who helped to explain zebra magnetic patterning at rifts. During the 1980s its ship towed a long line of hydrophones around the seas of Britain, obtaining higher-quality seismic profiles of the Earth below the ocean-bed than are possible from land. BIRPS' most fundamental discovery is the striking difference between the upper and lower crust. The upper crust is relatively blank, except for a few structures that dip, which can be regarded as faults. The crust below 7–8 miles (11–13 km) tends to be full of reflections, ending at the Moho. The pattern is repeated in western North America, northern France, parts of Germany and in Australia. The reasons are not yet understood.

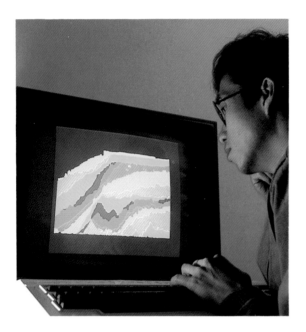

Computer graphics on a supercomputer are used for seismic sequence analysis during a search for land formations likely to contain oil. Techniques originally devised for oil exploration have been developed by scientists to probe the Earth's crust and mantle, and begin to create maps of the inside of the planet.

Ocean structure—the way currents move large bodies of water over great distances—is easier to measure than crustal structure, but hardly less mysterious. We can observe, for example, courtesy of Landsat, that the Gulf Stream is not a broad river but a meandering one with whirls and eddies. One of these will occasionally totally enclose a pocket of either colder or warmer water; fish will remain in this natural revolving swimming pool and do not migrate to the surrounding water. And off the western coast of South America, where an upwelling of deep, cold water gives rise to one of the world's richest fisheries, there periodically strikes the dread El Niño (the Child). Then the Peruvian fish catch fails because the upwelling of cold water has been blanketed by a great mass of invading warm water from the north, as Landsat images show.

Holes in the winter ice around Antarctica, known as polynyas, show how complex is the movement of the oceans. Polynyas have been extensively observed by microwave sensors on satellites. Microwaves, rather than visible light or infrared, are well suited to distinguish ice and water, because the emissivity of each is very different at certain microwavelengths. The polynyas come in two varieties: coastal, created by wind, and open-ocean. A very large open-ocean one, measuring 220×625 miles (350×1010 km), was observed in the Weddell Sea over three years, 1974–6, disappearing with the summer and reappearing in the same place each winter, as if it remembered its position. In 1977 it did not reappear and has not been seen since.

The cause of such localized holes in the ice far from land must presumably be an upwelling of warm water, preventing ice from forming. This warm water is thought to have travelled in deep currents from the equatorial regions. Rising to the surface in the polynya, it cools, approaching equilibrium with the temperature and composition of the atmosphere, before sinking again. The deep ocean is thus ventilated, establishing a rough balance between oceanic concentrations of dissolved gases such as carbon dioxide and the levels of those gases in the atmosphere. Understanding polynyas could therefore help us to understand the greenhouse effect caused by the addition of unparalleled quantities of carbon dioxide to the atmosphere. How influential are polynyas in the overturning and ventilation of the Southern Ocean? We do not yet know, but the phenomenon may provide a clue to the worrying process of global warming.

Urgent study of the hole in the ozone layer over the Antarctic has produced a clearer picture of atmospheric processes in that region, though major uncertainties remain. Balloon-borne measurements of ozone concentration above Antarctica began in 1957 with the International Geophysical Year. In October 1984 they showed that during the spring there had been a dramatic fall in concentration, which was subsequently confirmed by satellite measurements made by NASA's Total Ozone Mapping Spectrometer on board its Nimbus 7 satellite. Such was the international concern that some 150 scientists and support personnel assembled at Punta Arenas in Chile to conduct the Airborne Antarctic Ozone Experiment. Between August and October 1987 an airborne laboratory measured the ozone concentration and related atmospheric chemistry.

The result strongly suggested, but did not prove, that chlorofluorocarbons (CFCs), the inert gases used as coolants in refrigerators and air conditioners and as propellants in sprays and agents for producing foams, are generating chlorine atoms that are catalyzing the destruction of ozone in the stratosphere of Antarctica during the spring. Polar stratospheric clouds (PSCs), a unique feature of the frigid Antarctic winter, appeared to be a possible agent assisting the reaction. But dynamic processes, such as wind, that redistribute ozone in the atmosphere without destroying it, were also found to be important. In a field of research as new as atmospheric ozone chemistry, scientists could find no definite cause for the hole.

The evidence for global warming is less dramatic than that for the destruction of the ozone layer—so far. Maps showing small temperature rises all over the world, and

The uranium-rich San Rafael Swell in Utah, photographed by Landsat. *False-color processing of such images, combined with information from the ground, can sometimes be used to identify mineral-bearing rocks. Tin-bearing granite, for example, affects the vegetation growing upon it, and may reveal itself under satellite scanning and analysis.*

charts of small but increasing concentrations of carbon dioxide and other gases such as methane that act as atmospheric blankets, do not have the same public impact as the jagged "hole in the sky". An analysis of temperature records since 1860, carried out in 1988, shows that the average global temperature has increased by between 0.9 and 1.25°F (0.5 and 0.7°C), with the greatest increase taking place in the past decade. Atmospheric measurement shows that the annual net atmospheric gain in carbon, from burning of fossil fuels and deforestation, is about three billion tons, or 0.4 per cent. That these two facts are directly linked is almost universally accepted by scientists. It is when we contemplate the consequences of unchecked global warming that maps come into their own. Melting of the polar ice caps will flood large areas of coastline all over the world. Parts of low-lying cities such as Miami and Washington DC would be severely affected, and a recent map even shows water lapping at the steps of the White House. Parts of whole countries—the Ganges delta of Bangladesh and the Nile valley of Egypt, for example—would be under water. The Maldive islands could virtually vanish.

But maps are also a powerful tool to help governments overcome visions of disasters such as these. In the final analysis, maps are simply efficient ways to organize large amounts of related information visually; and that is a function that will become more and more in demand as we try to monitor and curb undesirable man-made activity, penetrate the processes that keep the Earth habitable, and mediate the interaction between them. "Geophysiology is [now] at the information-gathering stage, rather as was biology when Victorian scientists went forth to distant jungles to collect specimens," writes the distinguished but unorthodox scientist James Lovelock in his influential book about the Earth—or Gaia, as he calls her after the ancient Greeks' personification of Earth: *The Ages of Gaia: A biography of our living Earth*. Gaia is "a kind of geochemical myth for our time," according to *Scientific American*. Lovelock describes how it first glimmered for him while working at the Jet Propulsion Laboratory in Pasadena, California, in the late 1960s: "When I first saw Gaia in my mind I felt as an astronaut must have done as he stood on the Moon, gazing back at our home, the Earth. The feeling strengthens as theory and evidence come in to confirm the thought that the Earth may be a living organism."

Some people have taken this to mean that global catastrophe is impossible, that "Gaia will look after us." Lovelock disagrees. "If the concept means anything at all, Gaia will look after *herself*," he says. "In Gaia we are just another species, neither the owners nor the stewards of the planet. Our future depends much more upon a right relationship with Gaia than with the never-ending drama of human interest." Established scientists seem to be tending towards this view. The US in conjunction with many other nations, is launching a Mission to Planet Earth, "to collect data which will allow us to make predictions about what humans are doing to the Earth," in the words of Dr Lennard Fisk, one NASA scientist involved. "We have reached that point in human history where we are able to affect the global environment. We need to understand what that effect is in order to protect ourselves."

This project is the most far-reaching international scientific collaboration ever attempted. The data it will generate will exceed in just a few days all the data about the Earth previously accumulated by satellites and other methods of measurement. Only now do we have the computers and Geographic Information Systems powerful enough to cope. Inevitably, the project has its sceptics, who say that the Mission will produce too much too late, and ask how all the data will be handled and who will analyze it. But, as Fisk points out: "You have to ask, What's the alternative? These days we are in the business of protecting our global environment. The amount of carbon dioxide increase in the atmosphere, the methane increase in the atmosphere, deforestation, desertification, ozone depletion: these are real problems, not just things that people might imagine might happen. They are happening. And the question is, What are the consequences? There is an urgency in this project to determine those consequences, so that people can make sound policy decisions. Otherwise the consequences could be very detrimental." The shape of our world, the ghostly blue-green sphere that was first seen by the astronauts of Apollo 8, hangs in the balance.

Left *An ER-2 research plane at Stavanger in Norway prepares for Arctic atmospheric ozone research. Experiments in Antarctica in 1987, including ER-2 flights, demonstrated that chlorofluorocarbons (CFCs) are destroying ozone there. Now the same conclusions are being drawn at the North Pole. A worldwide Mission to Planet Earth, launched by NASA, is collecting data on every aspect of our relationship with the Earth.*

CHRONOLOGY

BC

c.2000 Sail first used on seagoing vessels, in Aegean.

c.1100 Phoenicians expand in Mediterranean region (to 700 BC).

c.900 Babylonian world map drawn on clay; first records of Babylonian astronomy.

c.750 Greek city-states begin to expand throughout Mediterranean.

c.610 Phoenician expedition sails round Africa (according to Herodotus).

c.530 Pythagoras, mathematician and mystic, active in Samos.

5th century Chinese astronomers begin continuous star observations.

380 First recorded description of Earth as sphere, by Plato in *Phaedo*, quoting Socrates.

334 Alexander the Great invades Asia Minor; conquers Egypt (332), Persia (330), reaches India (329); dies in Babylon (323).

350–200 Great period of Chinese thought: early scientific discoveries.

c.300 Pytheas, Greek navigator, possibly reaches Iceland.

c.310–230 Aristarchus postulates that Earth revolves around sun, according to Archimedes.

c.276– c.194 Eratosthenes calculates circumference of Earth.

221 Shi Huang-ti, of Ch'in dynasty, unites China (to 207).

214 Construction of Great Wall of China begins.

202 Han dynasty reunites China.

2nd century Possible use of lodestone in China.

c.190–c.120 Hipparchus, Greek astronomer: first to use latitude and longitude.

141 Wu-ti, Chinese emperor, expands Han power in eastern Asia.

c.138 Chang Ch'ien explores central Asia.

c.112 Opening of 'Silk Road' across central Asia links China to West.

c.64–c.21 Strabo, Greek traveller and geographer.

AD

23–79 Pliny the Elder, Roman scholar.

117 Death of Trajan; Roman Empire at its greatest extent.

127–145? Claudius Ptolemy mathematician, astronomer and cartographer, publishes his major works in Alexandria.

271 Magnetic compass in use in China.

Late 3rd century First Chinese silk maps with rectangular grid of co-ordinates.

c.350 Peutinger Table, Roman route map.

410 Visigoths invade Italy, sack Rome and overrun Spain.

6th century Cosmas Indicopleustes, traveller and Christian topographer.

632 Death of Mohammed; Arab expansion begins.

760 Arabs adopt Indian numerals and develop algebra and trigonometry.

800 Charlemagne crowned emperor in Rome: beginning of new Western (later Holy Roman) Empire.

c.1000 Vikings colonize Greenland and discover America (Vinland).

1095–1154 Roger II, Norman king of Sicily, establishes Christian and Muslim intellectual center, including mapmaker Al-Idrisi.

1137 Chinese 'Map of the Tracks of Yu'.

1206 Mongols under Genghis Khan begin conquest of Asia.

1275 Marco Polo arrives in China.

1280s Hereford *Mappa Mundi* drawn.

1291 Vivaldo brothers of Genoa attempt to sail round Africa to the East.

1375 *Catalan Atlas* completed by Abraham Cresques of Barcelona.

15th century Ptolemy's *Geography* revived in Europe during Renaissance.

1405 Chinese voyages in Indian Ocean begin (Cheng Ho).

1415 Portuguese capture Ceuta: beginning of Portugal's African empire.

1457 Map of the world drawn by Fra Mauro, Venetian monk.

1487 Bartholomew Dias rounds tip of Africa.

1492 Columbus discovers New World.

1494 Treaty of Tordesillas divides New World between Portugal and Spain.

1497 Cabot reaches Newfoundland.

1497/8 Vasco da Gama becomes first European to sail to India and back.

1498 Columbus discovers South America.

1499 Amerigo Vespucci, Italian navigator, reaches America.

1500 Cabral discovers Brazil.

1505 Portuguese establish trading centers in East Africa

1507 Waldseemüller map names New World after Amerigo Vespucci, not Columbus.

1513 Balboa sights Pacific Ocean.

1519 Cortés begins conquest of Aztec Empire.

1519–22 Magellan's ship circumnavigates globe.

1524–8 Verrazzano maps Atlantic coast of North America.

c.1525 Fernel first to measure degree of meridian, in France.

1532 Pizarro begins conquest of Inca Empire for Spain.

1533 First report of triangulation, by Frisius.

1538 World map by Mercator uses new projection.

1543 *Of the Revolution of Celestial Bodies* by Copernicus postulates sun-centered universe.

1566 Philip II commissions first detailed map of Spain.

1570 *Theatrum Orbis Terrarum* by Ortelius: first world atlas since Ptolemy's *Geography*.

1577–80 Drake circumnavigates globe.

1583 Ricci, Italian Jesuit, arrives in China and draws world maps for Emperor.

1584 Raleigh's expedition to colonize North America.

1607 First permanent English settlement in America, at Jamestown, Virginia.

1608 French colonists found Quebec.

1609 Telescope invented in Holland.

c.1610 Scientific revolution in Europe begins: Kepler (1571–1610), Bacon (1561–1626), Galileo (1564–1642), and Descartes (1596–1650).

1610 Galileo discovers Jupiter's moons.

1619 Batavia (Jakarta) founded by Dutch: beginning of Dutch Empire in East Indies.

1620 Puritans aboard *Mayflower* land in New England.

1640s Blaeu's *Atlas Major* published in twelve volumes.

1645 Tasman circumnavigates Australia and discovers New Zealand.

1652 Cape Colony founded by Dutch.

1667 Beginning of French expansion under Louis XIV and Jean-Baptiste Colbert: Paris Observatory founded.

1673 French scientific expedition to Cayenne suggests Newton's theory of gravitation is correct.

1675 Greenwich Observatory founded.

1687 *Principia Mathematica* by Newton.

1696 Completion of *planisphère terrestre*, scientifically compiled world map, at Paris Observatory.

1698/9 Halley's expedition maps magnetic variation in Atlantic.

1700 Jean-Dominique Cassini begins national survey of France.

1736–44 French Académie's expeditions to Lapland and Peru prove that Earth is flattened at Poles, as predicted by Newton in 1680s.

1762 John Harrison's No. 4 chronometer demonstrates accurate longitude determination.

1764 Mason-Dixon survey of Pennsylvania and Maryland begins.

1764 Rennell begins surveying Bengal Presidency.

1768–71 Cook's first voyage to Pacific.

1772–5 Cook's second voyage.

1775–83 American Revolution.

FURTHER READING

1777–9	Cook's third voyage.
1782	Rennell's first edition of *Map of Hindoostan*.
1787	Triangulation across English Channel links surveys of Britain and France.
1791	Beginning of British Ordnance Survey.
1793	Completion of French *Carte de Cassini*, first scientifically conducted national survey.
1802	Beginning of Great Trigonometrical Survey of India, under Lambton.
1803	Louisiana Purchase more than doubles size of USA.
1804–6	Lewis and Clark expedition crosses USA by land.
1810	Clark's map of American West.
1818	Britain becomes effective ruler of India, after defeating Marathas.
1838	Electric telegraph invented in Britain: soon used in longitude determination.
1842–8	Frémont's expeditions to American West.
1845–50	GTS triangulates Himalayan peaks, including Peak XV (Everest).
1848/9	California Gold Rush.
1848	*Map of Oregon and Upper California* by Preuss and Frémont: first accurate map of West.
1856	Peak XV in the Himalayas declared highest mountain in the world, named Mount Everest.
1860	California Geological Survey founded.
1872–6	World voyage of oceanographic vessel *Challenger*.
1879	US Geological Survey founded.
1909	Peary reaches North Pole.
1911	Amundsen reaches South Pole.
1912	Wegener publishes theory of Continental Drift.
1913	Greenwich meridian accepted internationally as prime meridian.
1947	Oceanographic vessel *Atlantis* discovers Mid-Atlantic Ridge.
1957	International Geophysical Year: scientific research stations established in Antarctica.
1957	First space satellite, *Sputnik 1*, launched by USSR.
1961	First man into space, Gagarin, launched by USSR.
1968	Earth first seen from space by US Apollo 8 mission.
Early 1970s	Mapping of Amazon Basin by airborne radar.
1972	*Landsat I* Launched by US.
1984	First observation of ozone 'hole' over Antarctica.
1987	Airborne Antarctic Ozone Experiment suggesting chlorofluorocarbons (CFCs) responsible for 'hole' leads to international action against CFCs.
1989	NASA launches Mission to Planet Earth.

The Age of Bede, trans. J. F. Webb and D. H. Farmer, London, 1965

The Age of Reconnaissance, J. H. Parry, London, 1963

The Ages of Gaia: A Biography of our Living Earth, James Lovelock, Oxford, 1988

The Amazon: The Story of a Great River, Robin Furneaux, London, 1969

Ancient Cosmologies, eds. Carmen Blacker and Michael Loewe, London, 1975

Before Columbus, Felipe Fernandez-Armesto, Basingstoke, 1987

A Brief History of Time, Stephen W. Hawking, London/New York, 1988

The Broken Spears, Miguel Leon Portilla, London, 1962

Captain James Cook, Alan J. Villiers, London, 1967

The Chinese View of their Place in the World, C. P. Fitzgerald, Oxford, 1962

The Conquest of America, Tzvetan Todorov, New York, 1982

Continental Drift, H. and M. P. Tarling, London, 1971

The Dawn of Modern Geography, 3 vols., C. R. Beazley, London, 1897

The Discoverers, Daniel J. Boorstin, London, 1984

The Discovery of North America, W. P. Cumming, R. A. Skelton and D. B. Quinn, London, 1971

The Discovery of the Sea, J. H. Parry, London, 1974

The Early Civilization of China, Yong Yap and Arthur Cotterell, London, 1975

An Encyclopaedist of the Dark Ages, Isidore of Seville, Ernest Brehaupt, New York, 1912

Espionage and Cartography in the Sixteenth Century, Corradino Astengo, 1986

The European Discovery of America: The Northern Voyages, Samuel Eliot Morison, New York, 1971

Exploration and Empire, William H. Goetzmann, New York, 1966

Explorers' Maps, R. A. Skelton, London, 1958

The Face of the Deep, Bruce C. Heezen and Charles D. Hollister, New York, 1971

The Fatal Impact, Alan Moorehead, London, 1966

The Fate of the Forest: Developers, Destroyers and Defenders of the Amazon, Susanna Hecht and Alexander Coburn, London, 1989

Geographical Lore at the Time of the Crusades, John K. Wright, New York, 1965

Geography in the Middle Ages, G. H. T. Kimble, London, 1938

The Great Temple of the Aztecs, Eduardo Matos Moctezuma, London, 1988

Greek and Roman Maps, O. A. W. Dilke, London, 1985

The Haven Finding Art, E. G. R. Taylor, London, 1956

Historical Records of the Survey of India, Reginald H. Phillimore, Dehra Dun, India, 1945–5

A History of Ancient Geography, E. H. Bunbury, London, 1879

History of Cartography, Leo Bagrow, revised by R. A. Skelton, London, 1964

The History of Cartography, Vol. 1, eds. J. B. Harley and David Woodward, Chicago, 1987

The Hole in the Sky, John Gribbin, London, 1988

Images of Earth, Peter Francis and Pat Jones, London, 1984

Indian Explorers in the Nineteenth Century, Indra S. Rawat, New Delhi, 1973

Isles of Gold, Hugh Cortazzi, John Weatherhill Inc, New York, 1983

The Journal of Lewis and Clark, ed. Bernard De Voto, Boston, 1953

The Mapmakers, John Noble Wilford, London, 1981

"Mapping Mount Everest: Surveying the Third Pole", Bradford Washburn, *National Geographic*, November, 1988

Maps and their Makers, G. R. Crone, London, 1953

The Marine Chronometer: Its History and Development, Rupert T. Gould, London, 1923

Memoirs of My Life, John Charles Frémont, Chicago, 1887

A Mountain in Tibet, Charles Allen, London, 1982

Never at Rest: A Biography of Isaac Newton, Richard S. Westfall, Cambridge, England, 1980

Philip II, Geoffrey Parker, London, 1979

Ptolemaic Alexandria, P. M. Fraser, Oxford, 1972

The Realm of Prester John, Robert Silverberg, New York, 1972

Science and Civilization in China, Joseph Needham, Cambridge, 1959

The Scientific Revolution, 1500–1800, A. R. Hall, London, 1956

Some Notable Surveyors and Map-Makers of the Sixteenth, Seventeenth and Eighteenth Centuries, George H. Fordham, Cambridge, Mass., 1929

The Story of Maps, Lloyd A. Brown, Boston, 1949

"The Structure of Mountain Ranges", Peter Molnar, *Scientific American*, July, 1986

Through a Glass Darkly, William Boelhower, Oxford, 1987

The World Encompassed, G. V. Scammell, London, 1981

INDEX

ACKNOWLEDGMENTS

The publishers would like to thank the following individuals and organiz-ations for their permission to reproduce the pictures in this book:
American Association for the Advancement of Science 181; Ancient Art and Architecture Collection 10, 11, 13, 21, 33, 50, 51, 52, 71, 78, 82; Ardea 175 (Grizzly); Ashmolean Museum, Oxford 20; Benson Collection, University of Texas 88 (map); Simon Berthon 17; Bibliothèque Nationale, Paris 69, 75 (Vasco da Gama), 113; Biblioteca Medicia Laurenziana 53; Beinecke Library, Yale 158–9; Bodleian Library, Oxford 86; Bridgeman Art Library 63, 85 (Montezuma), 118, 124, 156 (Napoleon); British Library 34–5, 38, 42, 44, 107, 108, 110, 114, 115, 129, 132, 134, 135, 136, 139 (map), 140 (theodolite), 142, 143 (Calcutta), 146; British Museum 13, 55, 59, 60, 62, 76, 92, 93; J. C. Ciancimino 46; Bruce Coleman Ltd 57 (double coconut), 111, 116, 126, 147, 153, 160 (Oregon trail), 164; Alex Connock 85 (Volodores); Allison Denyer 12, 16, 22, 28, 43, 61, 73, 75 (Belem Tower), 77, 98, 137, 139 (Survey tower), 148; Mary Evans Picture Library 29, 30, 56, 57 (General Gordon), 74, 83, 100, 106, 120, 121, 127, 143 (elephants); Gamma 175 (Yellowstone fire); Michael Holford 19, 32, 49, 72, 88 (Philip II), 89, 130, 140 (Tanjore); Hulton Picture Company 64, 133; Hutchison Library 24, 26, 37, 68, 177; Museo Naval, Madrid 80; NASA 170; National Archives, Washington DC 168; National Maritime Museum, London 45, 101, 119, 122, 123, 125; National Portrait Gallery, London 91; Newberry Library, Chicago 87; Peter Newark's Western Americana 95, 150, 156 (Jefferson), 160 (Bridger), 161, 162, 163, 165, 167; Rotterdam Maritime Museum 96–7, front and back endpapers; Science Photo Library frontispiece, 9, 18, 31, 102, 103, 109, 128, 173, 176, 179, 183, 184, 185; Space Frontiers Ltd 182; Survey of India 142, 146; Tate Gallery, London 104; Washington University, Washington DC 152; Woods Hole Oceanographic Institute 172.